SpringerBriefs in Systems Biology

For further volumes:
http://www.springer.com/series/10426

Sylvia M. Clay · Stephen S. Fong

Developing Biofuel Bioprocesses Using Systems and Synthetic Biology

 Springer

Sylvia M. Clay
Virginia Commonwealth University
Richmond, VA
USA

Stephen S. Fong
Virginia Commonwealth University
Richmond, VA
USA

ISSN 2193-4746 ISSN 2193-4754 (electronic)
ISBN 978-1-4614-5579-0 ISBN 978-1-4614-5580-6 (eBook)
DOI 10.1007/978-1-4614-5580-6
Springer New York Heidelberg Dordrecht London

Library of Congress Control Number: 2012947389

Printed on acid-free paper

Springer is part of Springer Science+Business Media (www.springer.com)

Preface

Different fields are often fraught with field-specific terminology and concepts that are loosely defined but used extensively by those active in that field. This is one of the barriers that makes it difficult for someone to learn about a new field of study. In the case of biofuels research and the biofuel industry, there are a vast number of disparate fields that apply to develop a biofuel process. Thus, for someone with interest in biofuels and wishing to become educated on relevant subjects, there may be a long initial period of deciphering terminology and linking concepts.

The intention of this Springer Brief is to provide a broad context of topics relevant to the development of biofuel processes. In particular, emphasis will be given to the recent fields of systems biology and synthetic biology as they relate to biological engineering. The content is meant to start by providing general introductory material into each of these fields and progress with more detail on concepts and methods culminating in highlighted research progress in systems biology and synthetic biology with relevance to biofuels.

Part of the broader context and content of this Springer Brief are focused on the general problem of scientific or technological decision making. While this may sound like an obvious and intuitive component that does not merit discussion, there remains a problem that biofuels impact so many different aspects that there are variety of different decisions to be made ranging from scientific research decisions to governmental policy decisions. Through this spectrum, any of the decisions can impact the speed and efficacy with which viable biofuel production processes may be developed and thus, it seems necessary to discuss explicitly some of the considerations that should be accounted for in developing a biofuel process.

Overall, we hope that this text will be useful to a broad range of readers and provide a broad sampling of material to provide a general perspective on the changing approaches to biological engineering while also providing sufficient examples to show relevance and progress in biofuel research.

Contents

Abbreviations

DMAPP Dimethylallyl pyrophosphate
DXP 1-Deoxy-D-xylulose 5-phosphate pathway
GHG Greenhouse gas
GPR Gene-protein-reaction
IMG Integrated microbial genomes
IPP Isoprenyl pyrophosphate
ISO International standards organization
KEGG Kyoto encyclopedia of genes and genomes
LCA Life cycle assessment
MEV Mevalonate pathway
TW Terawatt

Chapter 1
Introduction

Abstract This introductory chapter describes the broader historical context and perspective relevant to technology, engineering, and specifically application of technology and engineering to develop biofuel processes. Emphasis is given to the relationships between scientific research and global societal concerns with engineering acting as a bridge discipline between the two. Another point of emphasis that is highlighted is the need for thoughtful, integrative decision-making that accounts for aspects of research and societal concerns to arrive at processes that represent the confluence of the most desirable characteristics to meet society's needs.

Technology and engineering have long been connected to broad societal changes. The impact of technology and engineering on society easily goes back to the construction of simple tools in the Stone Age and progresses with metallurgy, the industrial revolution, the steam engine, and now advanced electronics, computers, and distributed information. Technological advances have helped to shape our society and have implications on our everyday lives.

Engineering is fundamentally a translational discipline that bridges scientific research and societal applications. While the goal is to develop technology that helps to address a problem or need in society, engineering progress is made by drawing upon advances in scientific research. Thus, in engineering there is a need to have a sound understanding of all the technical details of the relevant research and to balance that with a broad perspective of potentially competing societal interests (Fig 1.1).

As a gross overgeneralization, the majority of people on either end of the spectrum (research or society) may not have a great depth of knowledge about the other facet (e.g., a typical individual concerned about their energy usage likely is not knowledgeable about the depth and breadth of progress in alternative energy research). This concern is especially important for scientific researchers whose work drives technological innovation.

Individual researchers or research groups can be highly skilled and knowledgeable about their specific research area. As such, in terms of the research microcosm, research teams are very adept at formulating research plans and experiments

S. M. Clay and S. S. Fong, *Developing Biofuel Bioprocesses Using Systems and Synthetic Biology*, SpringerBriefs in Systems Biology, DOI: 10.1007/978-1-4614-5580-6_1, © The Author(s) 2013

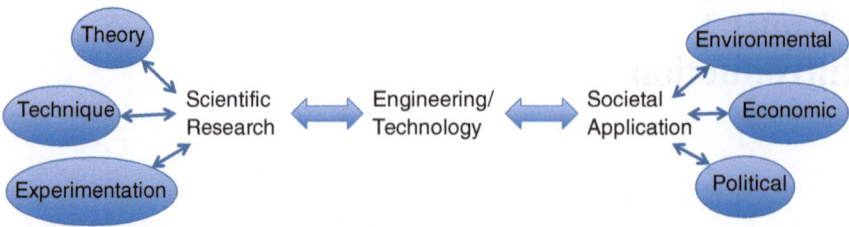

Fig. 1.1 Graphical depiction of the translational nature of engineering and technology to apply scientific research to address societal problems. *Ovals* depict illustrative examples of detailed concerns related to research or application

to advance knowledge in their area by balancing the state of knowledge in the field, techniques and methods available to researchers, and personal synthesis of ideas and hypotheses. As an academic pursuit, often the decision-making process for advancing research is largely dictated by the immediate environment of the research microcosm. Thus, not all research that advances our basic knowledge is easily translatable to an application.

One of the hallmarks of engineering and engineering education is being versed in problem solving by analysis, making appropriate assumptions, and reformulating problems (often in simplified forms). This is a decision-making process for technical problems. In the broader perspective of translating research to application and considering the variety of technical details from the research perspective and the global influences of the societal perspective, thoughtful decisions must be made, especially in terms of developing processes for biofuel production.

As a subject area, biofuels have a wide diversity of topics and details to consider. These range from social concerns that the general public is very aware of (food vs. fuel debate and economics of ethanol production) and technical research challenges in targeting the best starting materials, best fuel chemicals, and best organisms/processes for production. In addition to discussing these various topics, there is a need to consider all of these aspects to make decisions on where best to allocate resources and effort to quickly develop biofuel production processes with the best combination of beneficial characteristics. One main focus of this text will be to highlight some of this decision-making process.

In relation to biofuel production from a research perspective, biotechnology has historically utilized organisms for a variety of purposes including the early examples of alcohol fermentation and antibiotic production. A major move forward occurred with the identification and characterization of DNA as the fundamental building block of life coupled with molecular biology methods to replicate DNA in vitro (polymerase chain reaction). This led to directed modifications of organisms for various purposes including chemical production. The most recent major step forward in biological research is the improved knowledge of cell-wide components (genome, transcriptome, proteome, metabolome), their interactions (interactome), and methods to measure and manipulate whole cellular systems.

These recent knowledge and technology leaps have effectively moved biology to being an information-rich field, where early biology was characteristically information poor. With changes in available information, it may be appropriate to consider the decision-making process in biological research, especially for biofuels research where many competing constraints can be identified.

We believe that there exists sufficient knowledge from both a biological perspective and a chemical perspective to take a global view of biofuel research to suggest research avenues that would have the highest possibility of addressing all of the parameters needed for a successful fuel alternative. From a biological perspective, systems biology has helped to develop tools and a knowledge base to gain a broad, detailed perspective of cellular function and synthetic biology tools are facilitating design and controlled expression of different genes.

In relation to biofuel production from a societal perspective, there are a large number of different important and potentially competing interests to consider. These various competing interests include economic, environmental, political, and ethical considerations. In terms of developing biofuel processes, the various global societal considerations must be considered and may play a key role in identifying and developing the most promising biofuels to target and processes to produce them.

The remainder of the contents of this text is divided into Parts I, II, and III to consider the aspects of research perspectives, societal perspectives, and engineering to develop biofuel processes. Part I will provide background information on biofuels including some of the competing broader considerations for biofuel production. Part I also includes discussion of life cycle assessment (LCA) as a decision-making framework. Part II will provide information on the recent research areas of systems biology and synthetic biology that have direct relevance to bioprocessing and biofuel production. Part II will provide overviews of systems biology and synthetic biology including both experimental and computational techniques and provide example illustrations on how these apply to biofuel-producing bioprocesses. Part III will focus on describing the integration of information and approaches to engineer a biofuel-producing process and provide some perspectives on current and future areas of emphasis.

Part I
Societal Context

Part I
Societal Context

Chapter 2
Biofuel Context

Abstract Liquid transportation fuels are used globally on a daily basis at a high consumption rate that is projected to rapidly increase over the next several decades. With dwindling, finite supplies of oil-based transportation fuels, there is an urgent need for alternative fuels. However, most proposed alternative fuels and fuel production schemes have potential impacts on food supplies, the environment, or finances. There are a diversity of potential starting materials for biofuel production, target fuel compounds, and organisms to use for production. Determining the best combinations of input, output, and process to satisfy the fuel demand while addressing additional societal concerns could lead to a sustainable fuel source.

Around the world and particularly in the United States, liquid transportation fuels for use in vehicles with internal combustion engines are a dominant, everyday convenience or necessity. Rough estimations of gasoline consumption in the US give that approximately 388.6 million gallons of gasoline are consumed each day (U.S. Energy Information Administration–www.eia.gov) and with a total population of about 314 million people (U.S. Census Bureau–http://www.census.gov/population/www/popclockus.html) which means that on average each person in the US uses almost one and a quarter gallons of gasoline each day. This level of consumption is a direct reflection of our reliance and the impact of liquid transportation fuels.

If the liquid transportation fuels that are currently used were plentiful and sustainable sources existed, the prominent role of these fuels in everyday life would not be a concern. However, since our current oil-derived fuels do not have sustainable sources, alternative fuels are desirable and necessary long term. A real issue arises when proposed avenues for sourcing alternative fuels have tradeoffs in other areas of everyday concern such as food supplies or money. Due to these potentially competing interests it is important to evaluate the different facets of potential biofuels and to consider their impact on fuel supply and other areas.

S. M. Clay and S. S. Fong, *Developing Biofuel Bioprocesses Using Systems and Synthetic Biology*, SpringerBriefs in Systems Biology, DOI: 10.1007/978-1-4614-5580-6_2, © The Author(s) 2013

2.1 Biofuels in the Energy Landscape

Most people readily acknowledge the need to develop additional or alternative sources of energy. What is typically not realized is the urgency to develop these alternative energies due to the extent of our (US and the world) current and projected future energy consumption. Above was given an example of the gasoline consumption in the US. The current human global energy demand of approximately 14.9 terawatt (TW) is predicted to rise to 23.4 TW by 2030 (an increase of more than 50 % in a just a fifth of a century) (Hambourger et al. 2009). Similar consumption projections hold true for liquid transportation fuels. Due to the finite amount of oil, there is an immediate need to develop alternative fuel sources that will work in our current engines and be easily accessible to the population. Biofuels (gasoline alternatives and biodiesel) are growing in popularity but there is no clear single fuel substitute for use with our current infrastructure.

2.1.1 Current Biofuel Situation

The term biofuel is a broad umbrella term referring to a fuel that is derived from a biological starting material. As a general term, there are a vast number of different biofuels that have been proposed that vary in combinations of starting material and target fuel. The challenging decision is to determine the best process for efficient production of large quantities of fuel when choosing from different starting materials (corn, corn stover, switchgrass, sugar cane/beets, soybeans, vegetable oil, etc.) to potentially produce different end products (ethanol, propanol, butanol, isoprenes, diesel). Since the goal of developing alterative fuels is to increase the availability of fuels, care must be given to how much energy goes into a production process to yield more usable energy. The processing, separation, and purification of biofuels use up energy that has to be accounted for when evaluating energy-efficient fuel sources. While ethanol production from corn is one classic example of first-generation biofuels, it also does not have a very high net gain of energy.

Production of biofuel from algae is one option for using a low amount of input energy for producing and processing its biofuel. Growth of algae is relatively simple with minimal growth requirements and sunlight is a primary energy input due to photosynthetic capabilities. Furthermore, algae are easily harvested and processed since culture takes place in a liquid mixture. Harvesting, transportation, and processing of land-based plant material are not as easily accomplished.

Processes for production of biofuels that utilize other photosynthetic microorganisms (such as cyanobacteria) will offer the same benefit as algae if the efficiency of fuel production with minimal energy usage is comparable. Larger eukaryotic organisms also have more metabolic processes occurring simultaneously and therefore use more of their energies on undesirable products. Microbial production of biofuels can be engineered for a maximum production of the biofuel, which will optimize the conversion of the energy supplied to the desired product.

2.1.2 Starting Materials: Feedstocks

Energy is neither created nor destroyed and therefore we need to find the most efficient way to take existing energy on our planet and convert it into fuel that we can use to power our current and future lifestyle. On the molecular level, breaking carbon–carbon bonds are the main source of energy but it takes just as much energy to make them as to break them. Currently, we are tapping into a source of long-chain carbon bonds in the form of crude oil stored under the surface of the earth but the formation and storage of this substance happens very slowly over time and we will run out of this source sooner than later. We need to find a solution that creates carbon–carbon bonds continuously and does so in a way that outputs more fuel than we use to produce the fuel.

Currently, the most commonly considered starting material for a biofuel production process is plant biomass (lignocellulosic biomass). Lignocellulosic biomass consists of cellulose, hemicellulose, and lignin that are three types of sugar polymers. Cellulose is a chain of glucose, hemicellulose contains xylose, glucose, mannose, galactose, and arabinose in varying quantities depending on the organism, and lignin is a chain of phenylpropanoid units. In order to use the sugars in lignocellulosic biomass, these polymers need to be broken down into their monomers, and that can be done enzymatically or chemically. Cellulases are used to break down cellulose. The enzymatic process is generally preferred due to its mild conditions but the biomass does have to be pretreated by steam or dilute acid to be digestible by the cellulases.

A number of organisms natively use feedstocks that are continuously produced naturally to grow and form carbon–carbon bonds as a part of their natural functions. Different types of organisms use different energy sources. Lignocellulosic biomass from trees and other wood sources as well as simple sugars from plants both provide carbon–carbon bonds that can be broken down for energy which is a much more sustainable process than using fossil fuels, but it is still a roundabout way of using the sun to photosynthesize plant mass and then break down the plant mass to produce fuel. This is why using photosynthetic organisms that produce biofuel directly from the sun's energy has gained considerable attention as a more permanent long-term solution.

Different classes of organisms undergo different processes that have the capability of producing a wide range of carbon chains. A diversity of metabolic capabilities exists in photosynthetic and non-photosynthetic prokaryotes and eukaryotes that can result in production of carbon-containing compounds with chain lengths ranging from 2 to 24 carbons. Carbon bonds take a relatively high amount of energy to create and therefore organisms tend to only make the bonds that benefit their growth and survival. Because of this, ethanol, a two-carbon molecule, is one of the easiest targets for sustainable fuel production. With each additional carbon added onto the chain as in propanol or butanol, it gets much harder to produce the molecules in organisms because the organisms will choose thermodynamically easier paths to process its metabolites and will not favor the upstream, high energy-requiring reaction.

2.1.3 Target Fuel Compounds

Finally, recent advances in synthetic biology have enabled scientists to use more commonly used industrial or laboratory organisms such as *Escherichia coli*, *Corynebacterium glutamicum,* and *Saccharomyces cerevisiae* as platforms for metabolic engineering, expanding native functionality by both broadening substrate range and extending chemical production capabilities (Jarboe et al. 2010; Krivoruchko et al. 2011). What this translates to is that when selecting a desirable target chemical for production, we are not limited to only those compounds that are natively produced in an organism. If a desirable chemical is identified, it is often possible to design a method for biological production of that chemical. In representative cases, Liao et al. have engineered butanol production in *E. coli* by introducing genes from *Clostridium acetobutylicum* (Astumi et al. 2008) and using systems approaches to explore ways to improve butanol tolerance in *E. coli* (Senger and Papoutsakis 2008; Lee et al. 2008).

Thus, it is possible to consider a variety of different chemical compounds as possible biofuels based solely upon their chemical characteristics without initial regard to the feasibility of production. A number of these compounds have been identified and targeted as compounds that are either natively produced or can be engineered into organisms (Table 2.1).

Due to the diversity of organisms, starting materials, and potential target compounds, biofuel production processes take many different forms. While incremental research progress is being made in a large number of different biofuel processes, it is still not clearly defined as to what process combinations will prove most fruitful. It is now becoming possible to answer this question from a more global, top-down perspective by independently evaluating chemical characteristics, starting material characteristics, and organism traits. By identifying the most promising starting materials and target compounds, the most suitable organisms can be targeted as host bioprocessing platforms.

2.2 Broader Impacts of Biofuels

2.2.1 Effect on Food Supply

There is a lot of concern as to how biofuel production will impact the global fuel supply since many biofuel sources are grown as crops. Corn, sugar cane, switch grass, and soybeans are the main crop-based energy sources and out of all of them, corn is by far the largest competitor with possible food production. Because the

Table 2.1 Examples of potential target biofuel chemicals and organisms that have been shown to produce that chemical

Biofuel	Pathway	Examples of organisms
Ethanol	Native	Zymomonas mobilis
		Pichia stipitis
		Clostridium thermocellum
		Clostridium phytofermentans
		Saccharomyces cerevisiae
		Escherichia coli
	Imported	Corynebacterium glutamicum
Biobutanol	Native	Clostridium acetobutylicum
	Imported	Escherichia coli
		Saccharomyces cerevisiae
Lipid fuels	Native	Cyanobacteria and algae
		Yarrowia lipolytica
	Imported	Escherichia coli
Hydrogen	Native	Cyanobacteria and microalgae
	Imported	Escherichia coli
Higher alcohols and alkanes	Native	Vibrio furnissii
	Imported	Escherichia coli

corn produced for biofuel is engineered to optimize its starch content, it is inedible and is taking up farm land that could be used to produce corn for consumption. This same argument and concern holds true for almost all land-based plant matter that is considered as a starting material for biofuel production. The common argument is that if arable land is used, it would be better used for food purposes rather than fuel purposes.

Potential alternatives to land-based plants are the use of aquatic, photosynthetic organisms such as algae or cyanobacteria. Biofuel production from algae will not affect the food industry because it can be grown in diverse water systems and only has to have access to sunlight. Most energy options depend ultimately on the energy gained from the sun via photosynthesis to some extent. Therefore, in a continuous process, photosynthesis is usually the limiting process in energy acquisition. Solar energy, reaching the surface of the earth at a rate of approximately 120,000 TW, is a sustainable resource exceeding predicted human energy demands by >3 orders of magnitude. If solar energy can be concentrated and stored efficiently then it has the capacity to provide for future human energy needs (Bungay 2004). There is a large amount of research going into optimizing photosynthesis in many different organisms for the advancement in energy production avenues. As with many other organisms, the potential yield of algal biofuel is limited by the fundamental inefficiencies in the photosynthetic conversion of solar energy to biofuel. As of now our direct use of sunlight through solar panels is far exceeding that of indirect use through photosynthesis. Compared with synthetic solar panels with reported 30 % efficiencies, photosynthesis has a maximum efficiency of 8–10 % (Hambourger et al. 2009).

2.2.2 Environmental Impact: Greenhouse Gas Emissions

Global warming awareness has forced the fuel industry to reduce their carbon footprint and there is a heavy focus to minimize the carbon emission levels of perspective biofuels. Both the carbon produced for the production of the biofuel and the burning of the biofuel are considered in the overall emissions amount. This also takes into consideration the carbon that photosynthetic organisms absorb which can make their carbon emissions negative. General consensus to date suggests that from a greenhouse gas emissions standpoint, biofuels would be a better option than oil-based fuels.

2.2.3 Economic Feasibility

In addition to all the other considerations that can influence the choice of biofuel and biofuel production process, economics may be the most influential. Due to the size of the automotive and fuel industries, there are considerable investments made in infrastructure, production facilities, and even government policy. The more practical aspect of the economics is the cost of the final fuel product when considering production costs. Blending larger amounts of biofuels with gasoline while keeping the price of gas the same will ensure a path to sustainability and fuel independence.

2.2.4 Sustainability

All of the different considerations related to biofuels can be restated as an issue of sustainability. While there are numerous contexts and definitions for sustainability, the consistent general concept involves maintaining and preserving the current quality of life for current and future generations of humans within the context of our natural environment. The idea of sustainability is not limited to environmental impacts alone but is comprehensive in touching upon almost every salient aspect and concern that motivates the continued development of biofuels.

References

Astumi S et al (2008) Metabolic engineering of *Escherichia coli* for 1-butanol production. Metab Eng 10(6):305–311. doi:10.1016/j.ymben.2007.08.003

Bungay Henry R (2004) Confessions of a bioenergy advocate. Trends Biotechnol 22(2):67–71. doi:10.1016/j.tibtech

Hambourger M, Moore GF, Kramer DM, Gust D, Moore AL, Moore TA (2009) Biology and technology for photochemical fuel production. Chem Soc Rev 38(1):25–35. doi:10.1039/b800582

Jarboe LR et al (2010) Metabolic engineering for production of biorenewable fuels and chemicals: contributions of synthetic biology. J Biomed Biotechnol 2010:761042. doi:10.1155/2010/761042

Krivoruchko A, Siewers V, Nielsen J (2011) Opportunities for yeast metabolic engineering: Lessons from synthetic biology. Biotechnol J 6(3):262–276. doi:10.1002/biot.201000308

Lee J, Yun H, Feist AM, Palsson BO, Lee SY (2008) Genome-scale reconstruction and in silico analysis of the Clostridium acetobutylicum ATCC 824 metabolic network. Appl Microbiol Biotechnol 80(5):849–862. doi:10.1007/s00253-008-1654-4

Senger RS, Papoutsakis ET (2008) Genome-scale model for Clostridium acetobutylicum: Part I. Metabolic network resolution and analysis. Biotechnol Bioeng 101(5):1036–1052. doi:10.1002/bit.22010

Chapter 3
Life Cycle Assessment

Abstract Due to the diversity of impacts the fuel industry has on society, it is often difficult to comprehensively evaluate the benefits and drawbacks of any given fuel or fuel production process. This is especially true when the impacts influence disparate fields that have their own considerations and metrics that may not be compatible. The analysis framework of a life cycle assessment (LCA) provides a comprehensive methodology for studying the combined consequences of different influences and has been implemented in many industries including biofuels. While LCA can be used to retrospectively analyze existing systems, it is also possible to use LCA to prospectively analyze a system to evaluate and determine the most promising choices to pursue. LCA is a potentially powerful component in helping to guide decision making for target biofuel research and development.

The fuel industry has direct impacts on many different aspects of society. This is especially true when considering biofuels where there are numerous considerations including economic, environmental, and social considerations. With the diversity of considerations affecting fuel production, distribution, and consumption it is often difficult to make decisions on which fuel or production scheme may be most suitable to meet certain criteria. One method for assembling and assessing different criteria to arrive at a decision is to conduct an life cycle assessment (LCA).

LCA is fundamentally a process for tabulating or collating information to facilitate analysis. As a process, there can be many different ways to conduct an LCA, but most commonly accepted and standardized method is the series of standards set by the International Standards Organization (ISO) (http://www.iso.org/iso/iso14000). A typical LCA includes the implementation of four major steps:

1. Define the system boundaries for assessment.
2. Life cycle inventory: create an inventory of inputs and outputs for the defined system. This step often includes determining a functional unit as a basis for evaluation.
3. Life cycle impact assessment. This step often includes normalization/weighting of inventory items.
4. Interpretation of results.

S. M. Clay and S. S. Fong, *Developing Biofuel Bioprocesses Using Systems and Synthetic Biology*, SpringerBriefs in Systems Biology, DOI: 10.1007/978-1-4614-5580-6_3, © The Author(s) 2013

The LCA analysis framework provides two main benefits, both of which are relevant to biofuel processes. The first benefit is that there is a requirement to consider and explicitly define the system that you will consider for analysis. This may seem like an intuitive and obvious step, but it has not always been given proper consideration in process development where there is often a technical impediment that causes perspectives to become narrowed to the single step in question. By considering the system definition, most LCAs ideally try to consider all aspects of a process from the raw materials and transportation to manufacture, consumption, and disposal. As such, LCAs have often been colloquially referred to as "cradle-to-grave" analyses.

The comprehensive view provided by a good LCA is meant to help to provide a broad, unbiased analysis of a process system that can be used for decision-making purposes. From a research perspective, increasing the chemical production yield using your favorite organism can be intellectually fruitful, but there is no guarantee that such research progress would translate to a process with commercial relevance. Engineering a strain of an organism with a 50 % increase in butanol production is great; however, if an organism is limited to using carbon sources that are expensive or available in limited quantities then it would be difficult to develop a commercial-scale bioprocess based on that organism. Another common biofuel-related example is the body of work focused on pretreatment of lignocellulosic biomass to hydrolyze cellulose and hemicellulose to hexoses and pentoses, respectively. In the pretreatment cases, a balance needs to be made between the severity (chemical and reaction parameters) of the pretreatment process and the output stream from the pretreatment process. Relatively harsh conditions can be used to hydrolyze the polymeric sugars into monomeric sugars that can be used for conversion to biofuels, but carryover chemicals or temperatures from the pretreatment process can be inhibitory to downstream processes. Thus, sometimes it may be necessary to consider a less severe pretreatment process that has lower yields if it interfaces with downstream bioprocesses better.

The second benefit of an LCA is that it is a generic analysis framework and information from diverse fields can be all considered within the same framework. As stated previously, biofuel production involves a number of different facets including technological, social, environmental, and economic. Each of these fields has its own terminology, considerations, and metrics. If considered individually, the exercise of comparing the impacts in each of these different fields would be the equivalent of comparing apples to oranges as numbers and values from one field do not necessarily translate comparably to a different field. The LCA framework is meant to facilitate this by defining a functional unit that can be used as a standard in each of the different fields. Thus, information from all relevant fields can be tabulated using a base functional unit and analysis can take into account impacts in different fields.

For biofuel processes, there are many different functional units that can be considered to be appropriate. The natural consideration is a gallon of fuel since we are all familiar with a gallon of gasoline (and the price of a gallon of gasoline at a gas station). The use of the volumetric quantity of a gallon is not necessarily the best to use however as it prohibits the comparison of different types of fuels that would have different energy content in a gallon. For fuels, the function that is being considered is the energy content or for transportation fuel applications the distance that can be traveled. If functional unit for an LCA is defined according to energy content, then all of the data that are tabulated, from environmental gas emissions to pricing, must be considered by the energy content functional unit. Then a comprehensive analysis can be accomplished that incorporates information from different fields in a coherent manner.

3.1 Life Cycle Assessment of Biofuels

Due to the large number of variables that can be involved with biofuel production, there exist a vast number (thousands) of different LCA studies. These different studies can vary in many aspects including the scope of the system being considered, and they most frequently differ in terms of starting material, process detail, and target fuel produced. Due to these variations, it is often difficult to directly compare the results of different LCA studies.

One relatively recent review article attempted to compare different published LCA studies to determine if any generalized results could be found (Cherubini and Strømman 2011). This study considered 97 different life cycle studies on biofuels. These studies included a broad spectrum of variables including the locale of the study, the input and outputs considered, processes involved, and scope of analysis. The definition of functional units for analysis also varied (input-related units, output-related units, agricultural land units, and year) again making it difficult to directly compare results. Despite these challenges in analysis, two general results were found. When considering a balance on greenhouse gas (GHG) emissions, biofuels resulted in a net reduction in GHG emissions when compared to fossil fuel-derived products. In addition, biofuel production schemes required a lower amount of fossil fuel input for production of fuels and thus resulted in a reduced consumption of fossil fuel.

For additional details on LCA studies of biofuels, there exist a large number of individual studies that can be considered and found through literature search engines. LCA studies can be conducted for different purposes but generally they can be conducted retrospectively on existing processes or prospectively on developing processes. Prospective use of LCA can be used as a decision-making tool as a framework to help guide research and process development efforts to identify the most promising avenues to pursue.

Reference

Cherubini F, Strømman AH (2011) Life cycle assessment of bioenergy systems: state of the art and future challenges. Bioresour Technol 102(2):437–451 (Epub 2010 Aug 6)

Part II
Research Context

Chapter 4
Systems Biology

Abstract Systems biology utilizes experimental and computational tools with a goal of understanding the intact, interconnected functionality of biological systems. The ability to comprehensively experimentally measure and computationally model all of the individual components in a cell has added new dimensions to our understanding of how cellular systems work. This knowledgebase provides the necessary foundation for modifying and engineering cellular function. Overviews of core experimental and computational systems biology methods are discussed and illustrations of application of these methodologies to biofuel research are provided.

At its core, systems biology ascribes to the premise that an important aspect of biological systems is the interaction and interconnectedness of the individual components of the system. Some have rightly contended that this is "just biology" but the major distinction for systems biology has occurred recently as both experimental and computational techniques have enabled biological systems to be studied as intact, integrated, functional systems. Technological and methodological improvements have enabled systems biology research and systems biology as a field has provided much of the knowledge base for biological engineering.

4.1 Experimental Systems Biology

The advent of experimental systems biology started in 1995 with the first complete genome sequence for a free-living organism (*Haemophilus influenza*) and gene expression microarrays built for a subset of *Arabidopsis thaliana* genes. Gene expression microarrays were quickly expanded for genome-wide studies as demonstrated in 1997 using *Saccharomyces cerevisiae*. These studies demonstrated the ability to take detailed experimental measurements for a given type of cellular component at the scale of an entire cell thus demonstrating the feasibility of studying cellular systems as intact entities.

Seeing these systems operate as a whole is the best possible scenario to examine the wild-type genes and any genetically modified states. Determining which

S. M. Clay and S. S. Fong, *Developing Biofuel Bioprocesses Using Systems and Synthetic Biology*, SpringerBriefs in Systems Biology, DOI: 10.1007/978-1-4614-5580-6_4, © The Author(s) 2013

genes are being expressed is fundamental in this exploration and provides a basis from which further experiments can be formed. Gene expression microarrays measure the mRNA transcripts present in the cell at any given time. To measure the mRNA, methods such as quantitative Northern blot, qPCR, qrt-PCR, short or long oligonucleotide arrays, cDNA arrays, EST sequencing, SAGE, MPSS, MS, or bead arrays may be used. Approaches to measure the proteins include quantitative Western blots, ELISA, 2D gels, gas or liquid chromatography, and mass spectrometry (MS). In either approach, getting expression data from the systems exposes phenotypes directly that may not have been predicted based on genotypic data. Also, once a baseline measurement of expression has been obtained, it is easy to track the changes in the cells after further perturbations.

Getting a complete picture of the system as it exists naturally helps to interrogate parts of the system that are not obviously or intuitively altered when the organism is studied in pieces. Seeing the less prominent characteristics gives us a more detailed description of the organism that is then used to make a more accurate computational model. Systems biology takes a look at larger systems instead of the individual processes that occur within a system. Traditional biology uses isolated processes to study cellular processes but potentially may miss properties that only arise when all of the other systems were functioning (emergent properties). Biological systems are so complex; it is difficult to believe that we could study one piece of it at a time and eventually know everything about it. Billions of years of evolution have enabled life to exist in amazingly organized and complex systems and we are only beginning to understand its intricacies. Systems biology views biological processes as a symphony where each part is important but you cannot experience the whole effect without all of the parts at work.

Studying complex systems involves an immense amount of data and there have been constant advancements in high-throughput characterization for vast data acquisition. High-throughput characterization has enabled biologists to create models of organisms and predict different phenotypes and behaviors under different conditions. Several other systems biology techniques include in silico modeling, omics (genome, transcriptome, proteome, metabolome, and fluxome), metagenomics, gene synthesis, and synthetic regulatory circuits, an enzyme and pathway engineering.

4.1.1 Core Experimental Methods

Since the first demonstration of system-wide measurements of DNA and mRNA transcripts, there have been an increasing number of technologies and techniques that can be considered part of systems biology. With the development of technological advances, the corresponding methodologies have used the suffix "omic" to specifically denote a system-wide measurement (see Box 4.1). This has led to the general term of "omic" data which would refer to a collection of system-level measurements. For our initial discussion, the focus will be on the core systems biology technologies as related to the central dogma of molecular biology (Fig. 4.1).

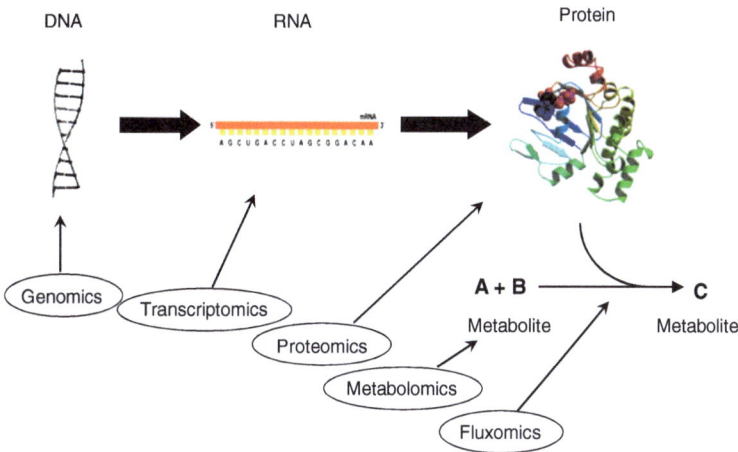

Fig. 4.1 Depiction of the central dogma of molecular biology and the relationship of different experimental systems biology measurements (shown in *ovals*) to each component of the central dogma

---

Box 4.1: Systems Biology Terminology

For the case of genetic content the applicable terms are genomic, genome, and genomics.

Adjective: genomic (i.e., genomic data)
Noun: genome (i.e., the genome of an organism)
Field: genomics (i.e., the study of genomes or genomic information)

4.1.1.1 Genomics

Given the centrality of DNA in biological processes, it was natural and essential to have systems-level measurements of DNA content for an organism as the foundation of systems biology. Fundamentally, this alludes to the grand challenge in biology of relating genotype to phenotype. If the genotype cannot be defined, then it is impossible to make a connection to function.

The standard for gene sequencing was developed in the early 1970s and it is referred to as the Sanger method. In this method, four different modified ddNTP's are fluorescently labeled and once attached to the DNA, they do not allow replication to continue and therefore are referred to as chain-terminating nucleotides. The lengths of the strands are determined and the ddNTP's indicate which of the four bases is present at each specific location. The main setback of this sequencing method is that each different ddNTP has to be run in a separate reaction and then the results have to be combined to get the entire sequence. Newer methods of sequencing are cutting back on the time required to obtain the sequence and the reliability of the results is increasing. Pyrosequencing is a more continuous process of sequencing where one base is added, and depending on how many bases are added to the sequence, a certain amount of light is emitted. Excess base is degraded and then a different base is added to continue the process. Illumina sequencing puts all of the bases in with the DNA and as the bases add to the strand they emit a color specific to the base and the colors are read in succession to give a read out of the sequence. While this method allows for more continuous sequencing, perhaps the most efficient method is nanopore sequencing. Electric current is flowing through the pore and anything that passes through the pore interrupts the current in a unique way because of its chemical structure. An entire strand of DNA can be passed through the nanopore and each nucleotide will disrupt the electric current in a unique way therefore giving a readout of the sequence.

4.1.1.2 Transcriptomics

To determine which genes are being expressed in an organism at any given time, a measure of the mRNA transcripts present or the "transcriptome" can be taken and it is an indicator of the variety and quantity of genes currently being transcribed. This information can be used to incorporate into a computational model that could then more accurately predict the behaviors of the organism under different conditions.

Differential display is a technique in which you compare two sets of mRNA from two samples in order to see altered gene expression from experimental variations. In order to compare the two samples they amplify both sets of mRNA with short arbitrary primers along with anchored oligo-dT primers (to bind only mRNA's with a poly-A tail) and compare the two results. So how do we obtain the transcriptome? RNA sequencing (RNAseq) is usually done by collecting all of the coding RNA strands and reverse transcribing them into cDNA and then using DNA sequencing as described above. Instead of this, the cDNA can be run on microarrays but there are issues with an overabundance of a few genes crowding the microarray chip and allowing some mRNA's to go unnoticed. So far the only techniques we have discussed are those that convert mRNA into cDNA, but this method has its limitations. This process involves ligating the mRNA which introduces biases and artifacts that could be avoided if the mRNA could be read directly, so direct RNAseq is now becoming available.

4.1.1.3 Proteomics

While the transcriptome can give you a good idea of what proteins can be present in an organism at any given time (presence of an mRNA transcript is a necessary but not sufficient condition for a functional protein), the number of mRNA transcripts does not directly correlate to the number of proteins. Not all mRNA transcripts are translated and transcriptomics does not account for post-translational modifications.

So instead of measuring the mRNA to estimate the number of proteins, you can actually measure them directly via proteomics. Quantifying the proteins gives confidence in knowing which processes are occurring in the organism under the current conditions, and how many of each protein are in existence at that time.

4.1.1.4 Metabolomics

While genomics, transcriptomics, and proteomics provide a wide range of information describing what is occurring inside of an organism, perhaps the most informative study is that of the new field of metabolomics. Metabolomics is the study of all of the metabolites within an organism, and metabolites are simply the molecules involved in metabolism. Knowing the levels that are present of each metabolite gives a detailed description of which cellular processes are occurring and at what rates. An elevated level of a certain metabolite in humans can indicate a certain disease or dysfunction and the same idea can be applied to microbes. To determine which metabolites are present and in what quantities, nuclear magnetic resonance (NMR) and many types of MS are most commonly used.

4.1.1.5 Fluxomics

Finally, to tie up the gap between metabolomics and proteomics, we have fluxomics. All of the reactions that occur in an organism are collected and given an input and output of metabolites per unit time. This is perhaps one of the most important datasets in terms of phenotypic characterization and it can be utilized in many different ways. As the model grows to contain most of the reactions possible in the organism, finite quantities can be used to predict a specific outcome. When a synthetic biologist is trying to produce a specific molecule such as a biofuel, they can engineer around the specific fluxes that lead to that molecule and focus on the most efficient pathways. This approach can even be taken to understand the fluxes that directly contribute to a cell's growth rate.

4.1.2 Progress for Biofuels

Experimental systems biology methods have been utilized for a variety of biofuel-related studies. The starting point for most systems biology approaches to biofuel production begins with genome sequencing. An organism's genomic content is the starting point for understanding the biochemical functionality of that organism. Progress on genome sequencing can be found in a number of different online databases including the GOLD database (www.genomesonline.org/). Having a genome sequence as a starting point is an almost necessary component for use of other systems biology experimental and computational tools. As such, there has been a focus on identifying and sequencing organisms of relevance to biofuel production. One example of this is the effort to sequence different members of *Clostridia* (Hemme et al. 2010).

After establishing a baseline of organism information by sequencing its genome, a number of additional systems biology methods can be implemented. With a genome sequence in hand, it is relatively fast to have a gene expression microarray constructed (or to conduct RNAseq) to allow for gene expression studies to be conducted. One of the applications of gene expression studies for biofuel applications is to study organism tolerance to a produced biofuel. Since most biofuels are toxic to cells, biofuel production schemes often have a physiological ceiling of biofuel concentration that can be attained before all of the production cells die. Use of gene expression arrays can help to identify and study the cellular toxicity of biofuels (Minty et al. 2011; Brynildsen and Liao 2009). Using gene expression analysis to study toxicity/tolerance can then be used to increase the solvent tolerance of a given strain. Examples of this include increasing the ethanol tolerance of *Escherichia coli* (Gonzalez et al. 2003), isobutanol tolerance in *E. coli* (Atsumi et al. 2010), and increasing the ethanol tolerance of *Saccharomyces cerevisiae* (Alper et al. 2006). Once understood, the cellular stress of biofuels can be reduced by amplification of tolerance-related proteins or addition of protectant in the culture media, which often increases the final titer of the target product.

In addition to genomics and transcriptomics, proteomic studies have often been used for biofuel research. One particular area of interest has been the study of different cellulase enzymes for the hydrolysis of lignocellulosic biomass into fermentable sugars. Cellulolytic organisms typically contain a variety of cellulase enzymes that utilize different mechanisms for cleaving polymeric sugars and the concentrations and mixtures of different cellulases has been of particular interest to study the most efficient methods for hydrolyzing cellulose and hemicellulose. Many of these studies have been conducted on specific organisms and aim to identify the cellulolytic proteins that are produced by individual organisms (Phillips et al. 2011; Adav et al. 2011). A unique attribute of some biofuel-relevant organisms is that cellulases that are produced to degrade lignocellulosic material are complexed into membrane-bound structures called cellulosomes. Proteomic analysis can help to identify the specific enzymatic composition of a cellulosome, as has been demonstrated for the cellulolytic bacterium *Clostridium thermocellum* (Gold and Martin 2007).

Proteomics, metabolomics, and fluxomics were combined in one study to gather data on *Chlamydomonas reinhardtii* to evaluate the growth rate and lipid production of these algae to evaluate if it was the best host for algal biodiesel production.

The first cyanobacteria was sequenced in 1996 and the strain was *Synechocystis* sp. PCC 6803 and as of 2011, 41 strains of cyanobacteria have been sequenced. Since they have such a small genome they are a perfect candidate for sequencing (Kaneko et al. 1996). Enough genomic and metabolomics data were gathered on this strain to also perform a flux balance analysis (Hong and Lee 2007). Obtaining a full metabolic analysis allows researchers to see where the carbon sources are being converted and can therefore direct them toward biofuel production. The transcriptome of *Cyanothece* 51142 was studied with microarrays to see the different functions of cyanobacteria in the light and dark cycles so that we can try to take advantage of one of the pathways for optimal production of biofuels (Stöckel et al. 2008).

Since algae are only recently being explored as an option for biofuels and other molecular-production purposes, it is also new to "omics". *Chlamydomonas reinhardtii* is one of the most studied strains of algae and its complete genome was not published until 2007 (Merchant et al. 2007). A transcriptome of *C. reinhardtii* is being built using a microarray that was developed specifically for that species. This microarray is being widely used and has been key to studying *C. reinhardtii's* light-regulated genes (Im et al. 2006) and its reactions to specific forms of oxidative stress (Ledford et al. 2007).

4.2 Computational Systems Biology

Biological systems (even the smallest ones) involve a large number of components. With the generation of large experimental systems biology datasets become a concurrent problem of developing methods to handle, process, compile, and analyze biological information. Due to the sheer size of some of the data types, efficient algorithmic approaches are necessary to even process raw data. The continuing challenges have been to develop standards in data formats that can be analyzed in an integrated manner and to develop analytical methods that provide useful insight into function.

The role and importance of computational methods to systems biology cannot be understated. As the technological capacity to experimentally measure more and more cellular components simultaneously has increased so too has the challenge in analyzing and interpreting results. For anyone who has received their first results/dataset from a transcriptional profiling experiment or a high-through-put genome re-sequencing study, the initial excitement of receiving the raw data quickly leads to a question of what to do with the data to figure out what it means. In studies that are conducted on a smaller scale, it is typical to have a clear hypothesis to be tested and proper controls implemented to directly test the idea in question. For system-level studies, this often is difficult as the control is typically an unperturbed, wild-type strain that can be compared to a strain that has been subjected to some perturbation. In these scenarios, many biological components can change simultaneously and it is challenging to determine if variations are due to a response to the perturbation or are a result of noise (intrinsic or extrinsic). The continued development of computational systems biology methods are meant to facilitate the analysis and interpretation challenge. Furthermore, as computational methods have improved, they have transitioned to being not only tools for post-processing data, but also tools for prospective design and prediction.

4.2.1 Core Computational Methods

To address the challenges faced with studying and analyzing intact, whole biological systems new analysis and interpretation methods have been developed. Two of the main areas these efforts have been focused on are: developing data formats and databases to standardize information and computational models to analyze and interpret data using simulations. Due to the large size of systems biology datasets, one of the main focal points and advantages of computational methods is to facilitate interpretation when it is difficult to intuitively understand the relationships between changes that can occur in hundreds or thousands of cellular components concurrently.

By establishing a suite of computational systems biology tools, researchers are able to approach scientific questions and engineering applications in a comprehensive, predictive manner. By standardizing and compiling information and data, it is easier to build an integrated, broad understanding of biological function. Coupling this with computational modeling allows for metabolic engineering applications such as biofuel production to be approached in a rational manner. Computational systems biology tools enable a shift from discovery-based science to rational design of engineering applications.

4.2.1.1 Data Management

Along with the excitement that occurred when the first systems-level experimental measurements were taken, there was a realization that data management and analysis would be a concern. Generally speaking, this was an information management issue and it largely gave rise to bioinformatics, but the problem can be stated much more simply. When the raw data output from a transcriptional profiling study results in relative numeric values for thousands of genes, what do you do with this data? For a simple case of studying gene expression changes for *Escherichia coli* between two conditions (run in triplicate), you would have a data output that has six columns of numbers and more than 4,400 rows. The challenge is to not only analyze and interpret this dataset, but it can be compared with other similar studies conducted in *E. coli* or different organisms.

While aspects of data analysis are unquestionably important to this field, the breadth of algorithms and approaches taken is large enough to warrant dedicated books/courses on their own. Here, we will highlight some of the logistical aspects of data management.

The primary resources for compiled biological information typically reside in Web-based databases. Just as the Internet has collectively become the main repository and access point for information in almost all fields, this has held true in biological research as well. The most comprehensive repositories of biological information now reside on the Internet as databases or data repositories.

There are a number of different databases and while not all of them will be listed here, some of the more commonly used databases will be highlighted.

One of the most widely accessed databases of biological information is the Kyoto Encyclopedia of Genes and Genomes (KEGG). This database is used as one of the central resources for gene and biochemical pathway content for specific organisms. Another commonly used resource for gene and pathway information is BioCYC (and all of its derivative organism-specific pages). These sources are centralized databases that provide organism-specific genetic and biochemical information and thus provide an initial link between genetics and metabolism (biochemical function).

Other databases specialize in providing more detailed information on different biological components (genes, proteins, etc.). For example, the Universal Protein Resource, called UniProt (www.uniprot.org), is a centralized database that provides details on proteins including sequence and functional information. Expert Protein Analysis System (ExPASy, www.expasy.org) was originally developed to as a Web-based tool to help to analyze protein sequences and structures. It has since expanded to include broader suites of bioinformatic analyses. Databases also exist for other aspects of biological systems such as the BRENDA database (www.brenda-enzymes.info) that is a central resource for enzyme information. The LIGAND database (www.genome.jp/kegg/ligand.html) contains information on chemicals and reactions.

While the previously mentioned databases primarily focus on compilation and dissemination of different facets of biological knowledge, other databases serve as central repositories for data. One such database is the GEO database that was used as a central repository for gene expression data. The National Institutes of Health (NIH) hosts one of the main centralized repositories for DNA sequence information in the National Center for Biotechnology Information (NCBI, www.ncbi.nlm.nih.gov/guide/). Databases such as the BIGG database (http://bigg.ucsd.edu) and ModelSEED (http://blog.theseed.org/model_seed/) serve as repositories for different biological models that have been developed.

Regardless of the information or data that is deposited in a database, one of the persistent problems has been establishing standardized formats for disseminating information. This requirement is necessary for allowing work to be conducted and shared between different researchers. For experimental data, there have been recent discussions and proposals for establishing minimum reporting features for all datasets that include standard data formats for specific data types and provenance information on the experimental parameters/setup associated with the dataset (Arkin 2008). The most developed of these data reporting formats is the Minimum Information About a Microarray Experiment (MIAME) format that was adopted for reporting gene expression datasets (www.mged.org/Workgroups/MIAME/miame.html).

An additional computational tool that has been useful to researchers is the development of visualization tools. Specifically, when considering system-wide measurements or analyses, it is often difficult to obtain a broad view changes or function within that system (for even small biological systems hundreds to thousands of components are simultaneously involved). When dealing with large datasets, the typical analysis involves the use of some statistical analysis with some arbitrary threshold/cut-off used to narrow down the number of variables to be interpreted. Given this analysis paradigm two problems immediately arise: (1) it is still often difficult to understand the collective functional effect of the subset of variables on the selected list and (2) the interpretation of what is functionally changing can vary greatly depending upon where the threshold/cut-off is set.

It is not possible to eliminate the inherent challenges of interpreting large datasets, but the practice of visualizing experimental data in an organized fashion often helps to provide a broader view of functional changes and often does not rely on establishing thresholds/cut-offs. Clustering algorithms have been one of the standard methods for organizing and visualizing data.

An additional useful method of visualizing data has been to integrate experimental data with biochemical pathway maps. A variety of tools have been developed to facilitate showing data on biochemical maps including: Cytoscape (www.cytoscape.org), CellDesigner (www.celldesigner.org), KEGG (using the KGML format, www.genome.jp/kegg). A main benefit of this method of visualizing data is that information is functionally organized by biochemical pathways. Thus, it is relatively fast and easy to identify areas of metabolism that are undergoing significant coordinated changes.

4.2.1.2 Computational Modeling

System-wide quantitative measurements of cellular components have enabled biological systems to be studied in a comprehensive manner. As mentioned above, one of the challenges has been determining methods for handling, disseminating, and interpreting the results of high throughput "omic" datasets. In the previous section, some of the issues with handling and disseminating information and data were briefly discussed. In this section, the focus will be on interpreting results from a systems biology perspective.

Computational modeling and mathematical modeling have traditionally been associated and implemented for disciplines where quantitative measurements can be made for critical parameters of interest. As biological research becomes increasingly quantitative the emphasis on computational or mathematical modeling has increased. This is particularly true in systems biology where large-scale quantitative measurements are made.

The concept of modeling in a biological context can be generalized as an approach to formulate and test hypotheses. Conceptual models descriptively depict a concept or process that is being studied such that predictions on behavior can be made and tested to assess the validity of the model (and by association the concept/process). Thus, models represent a structured framework for developing and testing biological hypotheses. As increasing amounts of detail are known, the complexity and detail of biological models has also grown.

One of the general tradeoffs that currently occurs in computational biological models is the balance between the scale of the model and the level of detail that can be achieved (Fig. 4.2). Some of the broad types of computational models that

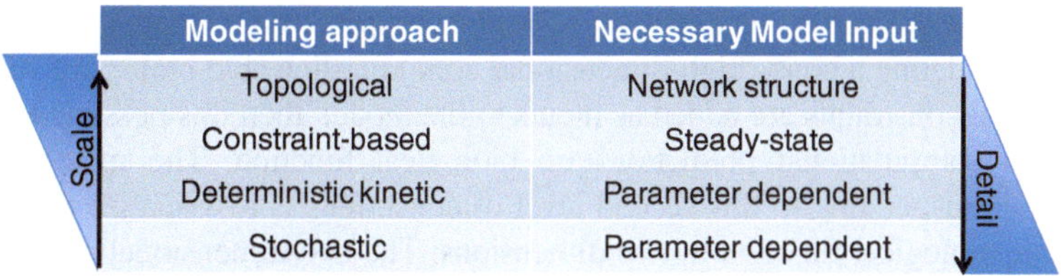

Modeling approach	Necessary Model Input
Topological	Network structure
Constraint-based	Steady-state
Deterministic kinetic	Parameter dependent
Stochastic	Parameter dependent

Scale — Detail

Fig. 4.2 Overview of modeling approaches commonly employed for biological systems depicting the tradeoff between the scale and detail of models developed by each approach

have been implemented for biological systems include topological models, constraint-based models, deterministic kinetic models, and stochastic models. When comparing these modeling approaches, topological models can be employed for any size system but contains the least amount of detailed information. Stochastic models would have the most detailed information (potentially accounting for biological noise), but currently cannot be implemented for large systems. When considering modeling approaches it is therefore important to consider the goal of the modeling and to choose an approach that best captures the desired attributes of the system.

One of the early comprehensive computational modeling projects was the E-Cell project that began in 1996 (www.e-cell.org/ecell). The first demonstration of this project was the construction of a self-sustaining computational virtual cell of *Mycoplasma genitalium* that included 127 genes. There have been subsequent releases of additional virtual cell models and concurrent development of software.

Another large-scale computational metabolic modeling approach, called constraint-based modeling, has been developed and implemented for a number of organisms with sequenced genomes. Biological systems are complex, interconnected, and specific functional properties change over time. These attributes of biology often make it difficult to develop large-scale computational models of biological systems. Constraint-based modeling is an approach that fundamentally develops models from a different perspective from deterministic kinetic models. While deterministic models identify, specify, and incorporate each individual parameter within the system to ideally run simulations resulting in discrete solutions, constraint-based models are flexible and underdetermined.

The fact that constraint-based biological models are flexible and underdetermined is a direct consequence of the underlying philosophy of constraint-based models. Constraint-based models are built by imposing high-confidence constraints on a system. The starting point for a constraint-based model is the assumption that the biological system being studied can achieve any imaginable function (the allowable solution space of function is unconstrained or infinite). Only by adding high-confidence constraints is the organism's function defined. The more constraints are added to the system, the more detailed (and hopefully accurate) the model becomes. The natural question is "What constitutes a high-confidence constraint?"

In the best scenarios, an initial set of high-confidence constraints can be implemented that define a cone (high-dimensional cone) of allowable biological function. If additional constraints are added as inputs to the model, then it is possible to obtain additional computational predictions on biological function. The most common analysis corresponding to this second level of constraints is to visualize the cone of allowable biological function in two dimensions. The two-dimensional projection of allowable function is called a phenotype phase plane and different regions of the phenotype phase plane depict different organism functional states. If an additional level of detailed constraints is used as input to the constraint-based model (i.e., experimentally measured oxygen uptake rates and substrate uptake rates) then it is possible to generate specific, numerical predictions of organism function (Fig. 4.3).

As currently implemented, the biological information that is most readily available on a large scale and that we have a high degree of confidence in is genomic and biochemical information. Genomic and biochemical information serve as the foundation for all of the current constraint-based biological models and are the first high-confidence constraints that are used. Typically, the combination of genomic and biochemical information for an organism is sufficient to establish the high-dimensional cone that defines the allowable, constrained solution space for an organism. There exist a number of review and methodology papers written that describe the process by which genomic and biochemical information are used to formulate a constraint-based model, so only a brief overview will be provided here.

If an organism has had its genome sequenced, the genome is annotated to identify or predict the genes that are present in the organism. The list of predicted, annotated genes is then correlated to the function associated with that gene. For genes that encode proteins with enzymatic activity, the associated function is the biochemical reaction that the protein facilitates. The stoichiometry of the biochemical reaction is then preserved and translated to a mathematical form where the convention is to have the stoichiometric coefficient of reactants noted with a negative sign and stoichiometric coefficients of products are given a positive sign. Every gene and every corresponding reaction for an organism is mathematically compiled using this convention. This association of gene to its protein product and corresponding biochemical reaction is called a gene-protein-reaction association (GPR). All biochemical reactions in an organism are represented in a mathematical form and the individual reactions are then compiled into a single mathematical matrix where each column of the matrix is an individual biochemical reaction and each row of the matrix is an individual metabolite. This tabulated matrix is called a stoichiometric matrix (it is compiled accounting of all biochemical stoichiometry) and is the core component of a constraint-based biological model. A schematic overview of this process is shown in Fig. 4.4.

The benefit of converting all of the biochemical reaction information into a mathematical representation in the form of the stoichiometric matrix is that matrix algebra and optimization algorithms can then be applied to study metabolic function of the system. The most commonly applied analysis of metabolism using a stoichiometric matrix is flux balance analysis (FBA) where a biological function (an objective function) is specified and linear optimization is run to predict the

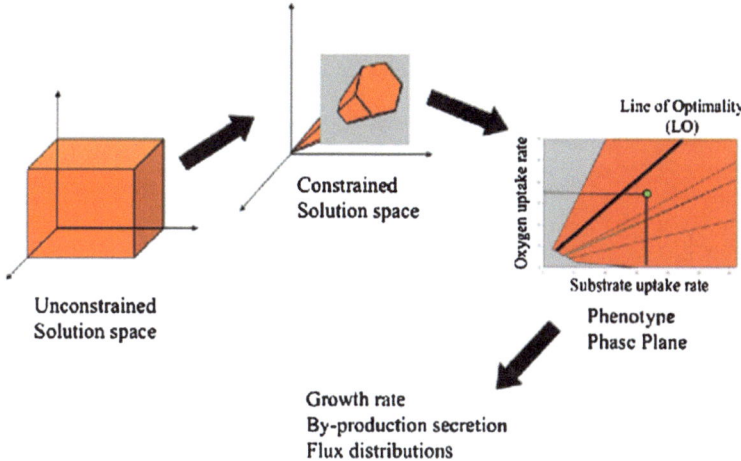

Fig. 4.3 Graphical illustration of the progressive application of constraints in constraint-based modeling to achieve progressively more detailed simulation results

maximum (or minimum) value for the specified biological function. An example of this would be to run a simulation to study the growth characteristics of an organism. In this case, cellular growth is the objective function and it is computationally implemented by formulating a reaction that chemically defines all of the requirements for cellular growth. Linear optimization is run and the output is a quantitative prediction of the expected growth rate of the organism that includes details of the specific biochemical reactions that are used to achieve the predicted growth rate.

Constraint-based metabolic models provide a readily scalable method for translating well-known, high-confidence genomic and biochemical data into a computational model that can predict cellular function. Related to the challenges of data interpretation and analysis mentioned in the previous section, constraint-based models provide a comparison with any available experimental data for a system. These attributes of a constraint-based model are generally applicable and beneficial for studying biological systems, but there are distinct advantages for using constraint-based models for metabolic engineering applications such as biofuel production.

The framework and content of a constraint-based model provide a computational method for analyzing and predicting strain designs for metabolic engineering applications. In the development of the stoichiometric matrix, all of the model content has specified GPR associations. This allows gene deletion or gene addition studies to be rapidly conducted. The functional effects of a gene deletion can be simulated quickly by removing the column in the stoichiometric matrix for the biochemical pathway(s) that are associated with the targeted gene. The effects of a gene addition can be simulated by adding a new column to the stoichiometric matrix to represent the biochemical function of the added gene. Gene deletion or addition simulations can be run manually or in an automated batch fashion and

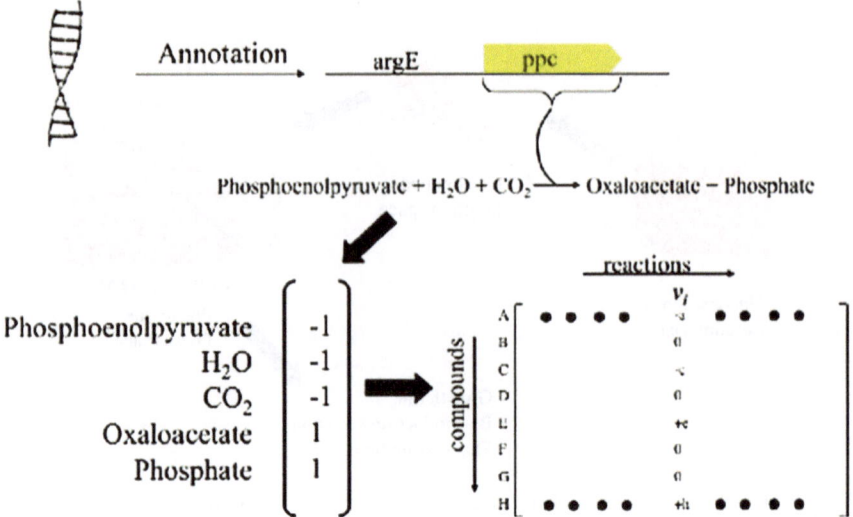

Fig. 4.4 Schematic depiction of the process used to develop a stoichiometric matrix for a constraint-based metabolic model

any combination of deletions/additions can be quickly studied. Furthermore, there are a growing number of algorithms that can search through all possible genetic combinations to propose the best strain designs for the production of a given metabolite (Ranganathan et al. 2010).

4.2.2 Progress for Biofuels

Computational systems biology methods provide foundational tools for knowledge and analysis of biological systems. For biofuels research, this is critical to advancing the development of biofuels research as many of the organisms and processes associated with biofuel production are relatively novel. Progress in both aspects of data management/dissemination and computational modeling have been made as related to biofuels.

In addition to the different general biological databases that have been developed for the collection and distribution of data there is a biofuel-centric database that is being developed by the U.S. Department of Energy. This database is called the Department of Energy Systems Biology Knowledgebase (Kbase, http://genomicsgtl.energy.gov/compbio/genomicscience.energy.gov/compbio/). The DoE Kbase will support the various aspects of computational systems biology discussed here including being a central repository for data generated that is related to biofuel production by bioprocessing. In addition, the DoE Kbase supports the development of new computational models and algorithms that will facilitate and support biofuel research. The development of the DoE Kbase is an ongoing activity that is fostered by specific government funding mechanisms to encourage research progress in this area.

In terms of computational modeling, a number of models have been built to support biofuels research. In particular, a number of constraint-based metabolic models have been constructed and analyzed for organisms with high biofuel production potential. Some of the best characterized constraint-based models related to biofuels are for organisms that have been used for the production of ethanol such as the ethanol-fermenting yeast *Saccharomyces cerevisiae* and high-ethanol producing bacterium *Zymomonas mobilis*. A potentially promising organism that has been modeled to produce ethanol or hydrogen is the anaerobic bacterium *Clostridium thermocellum*. *Clostridium acetobutylicum* that natively produces butanol during its solventogenic growth phase, the bacterium *Clostridium beijerinckii* which can utilize gaseous forms of carbon as an input and photosynthetic organisms such as cyanobacteria (*Synechococcus elongatus* and *Synechocystis*) and the microalgae *Chlamydomonas reinhardtii* have also been modeled. By growing the number of biofuel-related organisms that can be computationally studied, it may become increasingly feasible to conduct computational studies to predict the best strategies to implement to achieve high yields of organism-produced biofuel.

References

Adav SS, Ravindran A, Chao LT, Tan L, Singh S, Sze SK (2011) Proteomic analysis of pH and strains dependent protein secretion of Trichoderma reesei. J Proteome Res 10(10):4579–4596. doi:10.1021/pr200416t

Alper H, Moxley J, Nevoigt E, Fink GR, Stephanopoulos G (2006) Engineering yeast transcription machinery for improved ethanol tolerance and production. Science 314(5805):1565–1568. doi:10.1126/science.1131969

Arkin A (2008) Setting the standard in synthetic biology. Nat Biotechnol 26(7):771–774. doi:10.1038/nbt0708-771

Atsumi S, Wu TY, Machado IM, Huang WC, Chen PY, Pellegrini M, Liao JC (2010) Evolution, genomic analysis, and reconstruction of isobutanol tolerance in Escherichia coli. Mol Syst Biol 6:449. doi:10.1038/msb.2010.98

Brynildsen MP, Liao JC (2009) An integrated network approach identifies the isobutanol response network of Escherichia coli. Mol Syst Biol 5:277. doi:10.1038/msb.2009.34

Gold ND, Martin VJ (2007) Global view of the Clostridium thermocellum cellulosome revealed by quantitative proteomic analysis. J Bacteriol 189(19):6787–6795. doi:10.1128/JB.00882-07

Gonzalez R, Tao H, Purvis JE, York SW, Shanmugam KT, Ingram LO (2003) Gene array-based identification of changes that contribute to ethanol tolerance in ethanologenic Escherichia coli: comparison of KO11 (parent) to LY01 (resistant mutant). Biotechnol Prog 2:612–623. doi:10.1021/bp025658q

Hemme CL et al (2010) Sequencing of multiple clostridial genomes related to biomass conversion and biofuel production. J Bacteriol 192(24):6494–6496. doi:10.1128/JB.01064-10

Hong SJ, Lee CG (2007) Evaluation of central metabolism based on a genomic database of Synechocystis PCC6803. Biotechnol Bioprocess Eng 12:165–173. doi:10.1007/BF03028644

Im CS, Eberhard S, Huang K, Beck CF, Grossman AR (2006) Phototropin involvement in the expression of genes encoding chlorophyll and carotenoid biosynthesis enzymes and LHC apoproteins in Chlamydomonas reinhardtii. Plant J 48(1):1–16. doi:10.1111/j.1365-313X.2006.02852.x

Kaneko T et al (1996) Sequence analysis of the genome of the unicellular cyanobacterium Synechocystis sp. strain PCC6803. II. Sequence determination of the entire genome and assignment of potential protein-coding regions. DNA Res 3(3):109–136. doi:10.1093/dnares/3.3.109

Ledford HK, Chin BL, Niyogi KK (2007) Acclimation to singlet oxygen stress in Chlamydomonas reinhardtii. Eukaryot Cell 6(6):919–930. doi:10.1128/EC.00207-06

Merchant et al (2007) The Chlamydomonas genome reveals the evolution of key animal and plant functions. Science 318(5848):245–250. doi:10.1126/science.1143609

Minty JJ, Lesnefsky AA, Lin F, Chen Y, Zaroff TA, Veloso AB, Xie B, McConnell CA, Ward RJ, Schwartz DR, Rouillard JM, Gao Y, Gulari E, Lin XN (2011) Evolution combined with genomic study elucidates genetic bases of isobutanol tolerance in Escherichia coli. Microb Cell Fact 10:18. doi:10.1186/1475-2859-10-18

Phillips CM, Iavarone AT, Marletta MA (2011) Quantitative proteomic approach for cellulose degradation by Neurospora crassa. J Proteome Res 10(9):4177–4185. doi:10.1021/pr200329b

Ranganathan S, Suthers PF, Maranas CD (2010) OptForce: an optimization procedure for identifying all genetic manipulations leading to targeted overproductions. PLoS Comput Biol 6(4):e1000744. doi:10.1371/journal.pcbi.1000744

Stöckel J, Welsh EA, Liberton M, Kunnvakkam R, Aurora R, Pakrasi HB (2008) Global transcriptomic analysis of Cyanothece 51142 reveals robust diurnal oscillation of central metabolic processes. Proc Natl Acad Sci U S A 105(16):6156–6161. doi:10.1073/pnas

Chapter 5
Synthetic Biology

Abstract Synthetic biology started with an emphasis in experimental molecular biology through the demonstration that characterized DNA sequences which can be taken out of their native context and re-implemented in novel ways. The scope of synthetic biology research has rapidly increased with the improvement and development of tools for direct DNA synthesis and assembly of DNA molecules. These tools now make it possible to engineer biological systems precisely and accurately to reflect specific DNA-level designs. Application of synthetic biology techniques to biofuels research expands the scope of biological engineering that can be achieved where it is now possible to conceive, design, and implement large-scale changes to a cellular system.

While systems biology has provided a strong biological knowledge base for information and analysis, synthetic biology mainly focuses on tools and methods to manipulate or modify a biological system. A general goal of synthetic biology, which builds on advances in molecular and systems biology, is to expand the uses and applications of biology in the same way that chemical synthesis expanded the uses and applications of chemistry. Currently, synthetic biology has a main focus on nucleic acid methodologies (DNA, RNA) with one aim to provide standardized methods for genetic engineering.

Just as systems biology was enabled by technological developments, synthetic biology was also enabled by technology advances. Specifically, improved methods for DNA synthesis and molecular tools for assembling DNA are foundational to synthetic biology. Generally speaking, the improved ability to synthesize and construct DNA has led to the ability to more carefully interrogate genotype-phenotype relationships and also enabled the generation of novel genetically encoded biological function. Synthetic biology includes the design and construction of new biological entities, such as enzyme and even whole cells in order to create novel combinations of processes. The complexity of biological systems provides multiple types of machinery and a variety of options to include in those parts when engineering a system.

In its current form, synthetic biology operates primarily with nucleic acids. This means that design and implementation are done at a genetic base-by-base level. The most common type of sequence that is used is a gene where an average

S. M. Clay and S. S. Fong, *Developing Biofuel Bioprocesses Using Systems and Synthetic Biology*, SpringerBriefs in Systems Biology, DOI: 10.1007/978-1-4614-5580-6_5, © The Author(s) 2013

length is about 1,000 DNA bases. To facilitate communication at this level, synthetic biology has adopted several terms to describe different levels of organization (see Box 5.1). The term DNA "part" is used to refer to a standalone DNA sequence that has a discrete function. Parts can vary in length, but are typically of the order of tens to thousands of DNA bases in length. Representative parts could be a gene, a promoter, or a terminator.

If several parts are used in concert to achieve a more complicated function that collection of parts is termed a "device." The early demonstrations of the genetic toggle switch (Gardner et al. 2000) or repressilator (Elowitz and Leibler 2000) can be considered devices as well as the bacterial photography device (Levskaya et al. 2005). At a similar level of organization, the term genetic circuit is often used. Genetic circuits also typically incorporate a collection of DNA parts, but the difference in terminology is born from some of the early parallels to electrical engineering concepts that helped to lay the foundation for design. Some of the classic genetic circuits that have been constructed to date are biological equivalents to logic gates used in electrical circuitry (Wang et al. 2011; Zhan et al. 2010).

The host organism for implementing synthetic biology constructs is termed the "chassis." Currently, the most commonly used chasses in synthetic biology are *Escherichia coli* and *Saccharomyces cerevisiae* due to the large amount of information available for these organisms and the relative ease of working and genetically manipulating these organisms. In a generic sense, a chassis can be any system that contains all of the components necessary to functionally express a genetic construct, so it may be possible to engineer a biological chassis that is specialized for a given application. In the future, there may be a cellulolytic chassis that can be used as the starting point for biofuel applications that is specifically designed to efficiently breakdown lignocellulosic biomass and be streamlined for target fuel production.

Box 5.1: Synthetic Biology Terminology

Part: a single, relatively short DNA sequence with discrete, defined function
Device: a collection of multiple DNA sequences that integrates individual functions to achieve a novel coordinated function
Genetic circuit: a collection of multiple DNA sequences designed to operate as a functional circuit (design parallels to electrical engineering)
Chassis: host organism for implementing genetic constructs

5.1 Experimental Synthetic Biology

As with systems biology, the field of synthetic biology is not clearly defined (some would consider synthetic biology a natural progression of molecular biology), but there are some commonalities demonstrated by pioneering synthetic biology research. The earliest synthetic biology experiments (repressilator and toggle switch) utilized different genetic components found in various systems to conceptually and experimentally implement novel, controlled functions into a biological system. These early projects demonstrated some of the hallmarks of synthetic biology: novel design and utilization of genetic tools for experimental implementation. In parallel to a growing number of developed genetic circuits to demonstrate novel function, much of the work in experimental synthetic biology has been focused on developing methods for genetic engineering.

5.1.1 Core Experimental Methods

As the methods for DNA synthesis became more standardized, synthetic biologists took the opportunity to build a database of parts that could be combined in endless variations to build organisms with new functions. These DNA building blocks are called BioBrick parts and they are categorized into their different functions. There are BioBrick primers, ribosome binding sites, protein domains, protein coding sequences, terminators, and plasmid backbones. The database includes other various parts and combinations of existing parts as well. The ease of use with BioBrick parts comes from the systematic use of restriction enzymes specific to BioBrick parts, which makes assembling the DNA much like putting a puzzle together.

Different methods have been developed for assembling DNA fragments and BioBrick users can choose between 3A, Scarless, and Gibson assemblies. The name 3A refers to the three antibiotics used for selection with antibiotic resistance and it has the highest success rate with BioBrick parts. While there is no PCR or gel purification needed for this assembly, there is a scar left behind from the restriction and ligation process.

Additional methodologies have been developed to implement scarless assembly of DNA fragments. As implied by the name, these methods do not leave a scar from linkers. The absence of scars is very useful in assembling proteins and also allows the user to assemble parts that may not be compatible otherwise. Polymerase cycling assembly (PCA) runs similar to a PCR and uses oligonucleotides that all have flanking regions that combine to leave single-stranded gaps that a DNA polymerase then fills in. The DNA strands can be up to 50 base pairs and should overlap about 20 base pairs. Similar to PCA is another method called Isothermal Assembly (Gibson Assembly) where there is an overlap of about 20–40 base pairs and multiple strands of DNA can be joined in one reaction. Unlike PCA, Gibson is an isothermal assembly that occurs at 50 °C and runs for up to an hour making it one of the quickest assembly methods. In this method however the oligonucleotides contain the complete sequence; therefore, there is no need to fill in missing sections of the DNA although DNA polymerase is included in the reaction in case there are any gaps. This method becomes specifically useful when combining blunt-ended fragments. A T5 exonuclease is used to eliminate 20–40 base pairs from each end, leaving a single-stranded sticky end for ligation.

In order to make a recombinant gene more efficient it is important to make sure that the codons being used are easily used by the host. Codons are the three base pair sequences that code for a single amino acid and there are multiple codons that code for the same amino acid. Depending on what organism the recombinant genes are coming from or going to, the preferred codons will be slightly different. Codon optimization is the changing of a base pair in a codon in order to gain optimal production of the amino acid in the host organism. Optimizing the codons is most important when the recombinant DNA comes from a source that is genetically distant from the host organism such as plant DNA into bacteria. When optimized, it helps to improve improved translation rates, protein yields, and enzymatic activities.

Metabolic evolution provides a route for optimization and an option when determining the strongest strains. Allowing the organisms to compete for a food source allows nature to take over and the best strain can thrive and adapt and then evolve to be most suited for the given environment. Small mutations in the cell may happen naturally over time or they can be influenced by duplicating a gene using an enzyme with a high error rate. Allowing the cell to adapt makes the strain more stable and long lasting. Once the strain has evolved into a more robust state, the new genes can be used for the redesign of other systems.

5.1.2 Progress for Biofuels

While synthetic biology is a relatively young field, the global interest in biofuel research has led to the application of synthetic biology to several successful biofuel studies. One of the general approaches that are used is to use optimized heterologous expression of targeted genes to introduce novel biofuel production capabilities into an amenable host strain. Examples of this include the expression of different alcohol dehydrogenase genes from Saccharomyces cerevisiae and Lactococcus lactis in Escherichia coli to generate a strain of E. coli that produces isobutanol (Atsumi et al. 2010). Another demonstration was the engineering of the cyanobacterium Synecoccocus elongatus to produce isobutyraldehyde (Atsumi et al. 2009).

These demonstrations exhibit the ability to effectively use synthetic biology techniques to identify and express genes to modify specific pathways within an organism. This approach largely leaves the majority of an organism's biochemical network unaltered and intact. With the generation of an entire synthetic genome (Gibson et al. 2010) it may become possible to change the scale of synthetic biology engineering to include whole-cell design, not just pathway-specific design.

5.2 Computational Synthetic Biology

To complement and facilitate experimental synthetic biology research, computational methods are being developed. Due to the difference in system size and goals, the methods developed for synthetic biology are different from systems biology computational methods. Generally, the systems that are considered for synthetic biology are smaller in scale, but more detailed molecular-level dynamics are important.

5.2.1 Core Computational Methods

5.2.1.1 Genomic Information

One of the shifts associated with synthetic biology is the ability to explicitly control the base-by-base sequence of a genetic construct. With this level of control, it is possible to directly interrogate the effect of specific genetic changes to function. This is at the core of establishing genotype-phenotype relationships.

At one level, there is the need to compile and interpret sequence information. At a course-grain level, this is achieved during genome sequencing by genome annotation. There are several methods and pipelines that have been used to achieve genome annotation using computational means (though some input from experts is always beneficial). These include the Integrated Microbial Genomes (IMG) pipeline (http://img.jgi.doe.gov), the SEED system (www.theseed.org/wiki/Home_of_the_SEED), and a pipeline that is being developed through the National Institutes of Health (www.ncbi.nlm.nih.gov/genomes/static/Pipeline.html).

At a more detailed level, there are also programs such as GenoCAD (Wilson et al. 2011) that begins by developing a "grammar" for genetic parts. This approach considers genetic sequences as a language with specific rules that dictate the structure and function of different genetic parts. This grammar can be applied to not only studying DNA sequences for functional sequences, but can also be used as a basis for designing constructs to achieve new functional units.

5.2.1.2 Design Tools

A variety of tools are being established to help to design genetic circuits. Given the ability to experimentally construct any desired DNA sequence exactly, the design process has become truly open-ended. Any gene from any organism can be utilized in combination with any other gene. In a broader sense, even novel (previously undocumented) gene function can be proposed and tested.

Some of the design approaches that focus on utilizing existing biological information attempt to mine database information to propose a collection of genes (from any organism) to create a pathway that would achieve the desired goal. The "From Metabolite to Metabolite" tool demonstrates one iteration of this approach (http://fmm.mbc.nctu.edu.tw/). Using this tool, a user only needs to input a starting metabolite and a desired end metabolite. The algorithm then uses information from online databases such as KEGG, UniProt, and GeneBank to identify metabolic, protein, and sequence information, respectively. The output is a list of proposed pathways that could achieve the desired biochemical conversion from one metabolite to another.

Other tools have been developed to approach the design problem from a more generic approach. In these approaches, different methods represent chemicals/metabolites in a standardized form so that individual biochemical transformations can be considered in a stepwise fashion. Each proposed biochemical transformation can then be correlated to enzymes that would have the closest reaction mechanism (often as dictated by the enzyme commission number—EC#). This approach is conceptually similar to the old word game of changing one word to another by changing only one letter at a time while maintaining a valid word at each intermediate step (see Box 5.2). The different methods that have been implemented for this type of approach largely different on the method by which chemicals are represented (atomic mapping onto graph coordinates or linearized representation of atoms).

Box 5.2: Illustration of Stepwise Transformation

C A T
        Step 1: Conversion of "A" to "O"
C O T
        Step 2: Conversion of "C" to "D"
D O T
        Step 3: Conversion of "T" to "G"
D O G

5.2.1.3 Dynamic Simulation

With tools to study the basic information content of different DNA sequences and to propose different collections of genes to achieve a desired outcome, the final step that computational methods have addressed is the ability to simulate the function of the designed genetic circuit dynamically. A variety of different methods can be implemented to dynamically simulate small gene circuits including differential equation modeling, stochastic simulations, and agent-based modeling.

Differential equation models are the staple of dynamic simulations and can be implemented for small systems. Tools such as TinkerCell and SynBioSS can be used to develop computational models for small synthetic systems.

Stochastic simulations provide a simulation method that is different from differential equation models in which they are not deterministic and therefore account for some of the variability and noise that are inherent in biological systems. While this has the advantage of being a better representation of biological processes, there is often a tradeoff in terms of the size of the system that can be simulated and the computational time required to run simulations.

Agent-based models can be considered a subset of stochastic simulations, but with one major distinction. Agent-based models are formulated with discrete agents representing the physical entities within the system and thus it is possible to account for density and spatial effects. As with other stochastic simulations, there are limitations to the size of the system that can be studied largely due to computational resource limitations.

5.2.2 Progress for Biofuels

Currently, the number of studies linking computational modeling, synthetic biology, and biofuels is relatively limited. The majority of the work in this area has thus far focused on computational tools that can help with the design process, specifically in terms of helping identify non-native pathways and chemical targets that can be implemented using controlled heterologous gene expression.

As mentioned previously, one Web-based tool that can be used to help with pathway design independent of organism is the "From Metabolite to Metabolite" algorithm. Another recently developed algorithm used *Escherichia coli*, *Saccharomyces cerevisiae*, and *Cornyebacterium glutamicum* as host organisms and searched for non-native metabolites that could potentially be produced by heterologous expression (Chatsurachai et al. 2012). This algorithmic search was then coupled with flux balance analysis to determine feasibility.

Another common approach for studying the diversity of metabolites that can be produced by an organism via biochemical means is the use of graph theory or graph-based algorithms (Brunk et al. 2012). There exist different implementations of this approach to studying biochemical conversions. A recent implementation was used to specifically study the potential of different organisms to produce 1-butanol as a biofuel target (Ranganathan and Maranas 2010).

References

Atsumi S, Higashide W, Liao JC (2009) Direct photosynthetic recycling of carbon dioxide to isobutyraldehyde. Nat Biotechnol 27(12):1177–1180. doi:10.1038/nbt.1586

Atsumi S, Wu TY, Eckl EM, Hawkins SD, Buelter T, Liao JC (2010) Engineering the isobutanol biosynthetic pathway in Escherichia coli by comparison of three aldehyde reductase/alcohol dehydrogenase genes. Appl Microbiol Biotechnol 85(3):651–657. doi:10.1007/s00253-009-2085-6

Brunk E, Neri M, Tavernelli I, Hatzimanikatis V, Rothlisberger U (2012) Integrating computational methods to retrofit enzymes to synthetic pathways. Biotechnol Bioeng 109(2):572–582. doi:10.1002/bit.23334

Chatsurachai S, Furusawa C, Shimizu H (2012) An in silico platform for the design of heterologous pathways in nonnative metabolite production. BMC Bioinformatics 13(1):93. doi:10.1186/1471-2105-13-93

Elowitz MB, Leibler S (2000) A synthetic oscillatory network of transcriptional regulators. Nature 403(6767):335–338. doi:10.1038/35002125

Gardner TS, Cantor CR, Collins JJ (2000) Construction of a genetic toggle switch in Escherichia coli. Nature 403(6767):339–342. doi:10.1038/35002131

Gibson DG et al (2010) Creation of a bacterial cell controlled by a chemically synthesized genome. Science 329(5987):52–56. doi:10.1126/science.1190719

Levskaya A, Chevalier AA, Tabor JJ, Simpson ZB, Lavery LA, Levy M, Davidson EA, Scouras A, Ellington AD, Marcotte EM, Voigt CA (2005) Synthetic biology: engineering Escherichia coli to see light. Nature 438(7067):441–442. doi:10.1038/nature04405

Ranganathan S, Maranas CD (2010) Microbial 1-butanol production: Identification of non-native production routes and in silico engineering interventions. Biotechnol J 5(7):716–725. doi:10.1002/biot.201000171

Wang B, Kitney RI, Joly N, Buck M (2011) Engineering modular and orthogonal genetic logic gates for robust digital-like synthetic biology. Nat Commun 2:508. doi:10.1038/ncomms1516

Wilson ML, Hertzberg R, Adam L, Peccoud J (2011) A step-by-step introduction to rule-based design of synthetic genetic constructs using GenoCAD. Methods Enzymol 498:173–188. http://www.ncbi.nlm.nih.gov/pubmed/21601678. Accessed 13 July 2012

Zhan J, Ding B, Ma R, Ma X, Su X, Zhao Y, Liu Z, Wu J, Liu H (2010) Develop reusable and combinable designs for transcriptional logic gates. Mol Syst Biol 6:388

Part III
Developing Biofuel Processes
by Engineering

Chapter 6
Integrating Systems and Synthetic Biology

Abstract Research approaches to developing biofuel processes involve the integration of different aspects of biology, chemistry, and engineering. Recent developments in knowledge and technology have enabled a shift away from discovery-based, trial-and-error design to a more directed prospective design process. Systems biology and synthetic biology have contributed to this shift in methodologies in complementary ways. Systems biology provides much of the knowledge background and whole-cell modeling methods to enable cellular-level design. Synthetic biology provides DNA-level detail to design strategies and the experimental methods to directly implement proposed designs. Application of methodologies from these two fields provides a strong framework for cellular and molecular biological engineering.

Given recent research advances in systems biology and synthetic biology, new approaches to engineering biology can be taken (Fig. 6.1). Traditional approaches to metabolic engineering have largely been iterative discovery-based approaches. This approach starts with the discovery or characterization of an uncharacterized organism. The basic physiological and biochemical functions of the organisms are characterized and the naturally secreted metabolic end products are measured. From that point, a perturbation (environmental or genetic) is proposed and implemented. The effect of the proposed perturbation is then evaluated and iterative cycles of proposed perturbations and characterization occur to incrementally improve the desired function of the organism.

The discovery-based, iterative approach to biological engineering was necessary in part due to lack of knowledge and tools. It may now be possible to change the paradigm for biological engineering to a design-based strategy, where chemistry and biology knowledge inform a proposed design scheme. After implementation, the metabolic end products can be characterized and if successful, should closely match the predicted desired function.

The ability to achieve a design-based approach to biological engineering is enabled in large part by the developments in systems biology and synthetic biology. Research in systems biology has provided much of the necessary biological knowledge that is needed to understand a cellular system enough to attempt whole-cell

S. M. Clay and S. S. Fong, *Developing Biofuel Bioprocesses Using Systems and Synthetic Biology*, SpringerBriefs in Systems Biology, DOI: 10.1007/978-1-4614-5580-6_6, © The Author(s) 2013

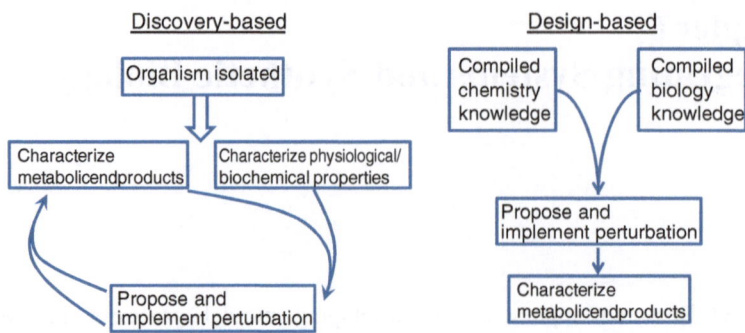

Fig. 6.1 Schematic of discovery-based and design-based approaches to biological engineering

design. In addition, computational biological models are being used to prospectively evaluate the effects of perturbations on cellular function. Synthetic biology progress has produced genetic engineering methodologies that enable almost any proposed genetic design to be directly implemented.

The ideal scenario for a metabolic engineering application such as production of biofuel is to be able to completely design all aspects of a biological system for optimal production in a similar fashion that blueprints or designs are made for other engineering disciplines. Starting from a blank slate and a given goal (biofuel production), a biological engineer would utilize any of the available tools (DNA in the form of genes and genetic parts) to design an optimal production system de novo. This approach would be accompanied by theoretical calculations on production yields and give a numerical basis for evaluating the success of the design.

6.1 Combining Biology and Chemistry

At the highest level of design considerations for developing a biofuel process is the need to make decisions that identify and select the best attributes of a chemical target and biological organism for production of that chemical. Thus, criteria must be established to evaluate the suitability of different chemicals as fuels. It is also necessary to establish a separate set of criteria to evaluate the suitability of different organisms as biofuel production hosts. The overlap between the chemical criteria and the biological criteria should provide an unbiased perspective to indicate promising possible design avenues.

From a biological perspective, a broad starting point of organisms and genetic information should be considered. From a chemical standpoint, a broad spectrum of high energy-content chemicals can be considered by employing Table 6.1. After the chemical selection is narrowed, each one should be evaluated for the best possible host organism (Table 6.2). By progressively applying filters/selection criteria to the biological starting point and the chemical starting point, the most

Table 6.1 Choose a list of possible biofuels and rate each one on a scale of 1–10 as to its suitability to each category

Chemicals	Toxicity	Enzymes available for synthesis	Sustainable Fuel Replacement	Score
Ethanol				
Propanol				
Butanol				
Isoprenoids				
polyketides				
Lactic acid				
Succinic acid				

Table 6.2 Once the target chemical has been chosen, choose a list of organisms to evaluate and rate each organism on a scale of 1–10 as to its suitability in each category

Organism	Existing Pathways	Existing Model	Genetically Tractable	Distance from metabolism	Score
Z. mobilis					
P. stipitis					
C. phytofermentans					
C. glutamicum					
C. acetobutylicum					
E. coli					
S. cerevisiae					
Cyanobacteria and algae					
Y. lipolytica					
V. furnissii					

bioprocesses can be postulated as a combination of biological organism and chemical product.

Determining the best combination of chemical target and biological organism likely will not result in a single, unique solution. As with all other aspects of biofuels research, it is inevitable that there will be some benefits and shortcomings for any given combination. Fortunately (or unfortunately), the scale of energy and fuel consumption that needs to be addressed means that it is also almost certain that multiple means of alternative fuel production are required.

Individual perspectives will ultimately determine the research path chosen, but there are likely to be some common considerations for anyone entering biofuel research. Some of these considerations from a biological side include an organism's ability to: utilize a sustainable input, be easily genetically modified, metabolically sustain diverse biochemical transformations, and be computationally modeled. From the chemical perspective, there are fuel attributes that need to be considered as well as toxic effects to cells, distance of the target chemical to existing biological pathways, and the availability of characterized enzymes to produce the target compound.

It is important to consider how far away possible target compounds are from native biological processes. One technique used to induce organisms to produce the desired compounds is to link the pathway to produce the target compound to the primary (or secondary) metabolism of the organism. In doing this, the organism will produce more of the target compound when it grows faster. By limiting the pathways used to only producing the desired compounds for growth and target compound production, the organism is able to use its resources more efficiently and more of the target compound is produced.

Choosing an organism that already performs as many processes as possible is ideal and then having enzymes from similar species available increases the chances of the recombination being successful. Enzymes have evolved to thrive in specific environments including temperature and pH and the specificity of these enzymes plays a role in recombination. More promiscuous enzymes can be chosen for recombination, but the recombination is less likely to be successful and the reaction is less likely to be efficient at the intended reaction mechanism. Enzymes that come from similar organisms as the host organism are more likely to thrive in similar environments and perform the desired reactions efficiently because years of evolution have optimized its performance.

How close the molecule is to metabolism is a factor that will determine production rate. The steps beyond known metabolite structure should be more basic chemical reactions with lower activation energies. Each additional reaction needed takes more time and energy that is why minimizing the additional reactions branching off a primary function of the cell is beneficial. Having pathways adjacent to the desired pathway that can be deleted will increase optimization opportunities. The metabolites that would have been used in the competing pathways are then used as inputs to the preferred pathway which leads to a greater flux of the reaction.

Both aspects of chemical toxicity and the ability to natively use sustainable inputs are important considerations as these attributes are often difficult to engineer into a cell if the cell does not natively have that capability. Chemical toxicity often triggers a systemic response involving a wide range of genes and physiological responses. For an organism that does not have native tolerance to a target chemical, redesigning the cell to have tolerance is a non-trivial endeavor. The same holds true for organisms that have limited abilities to utilize sustainable substrates. The two most sustainable source materials for organisms are either sunlight or lignocellulosic biomass. For organisms that do not natively have the ability to use these sources, it would be difficult to engineer photosynthetic or cellulolytic systems as both systems involve the interaction of a large number of genes.

The ability to model an organism or have a detailed knowledge of existing biochemical pathways is important for the overall design process. As a baseline for whole-cell design, a reasonable understanding of basal metabolism is necessary. If an organism is not well characterized in terms of its basic metabolic function, it is difficult to design novel functions because there is no way of predicting the interactions that would occur. Even if two organisms contain roughly the same genes and biochemical capabilities, there is no guarantee that they utilize these components in the same manner. This has been observed between closely related *Clostridia* species that are of interest for biofuel production. The ability to develop computational models for an organism may not be necessary, but is a step that greatly facilitates design by allowing fast and thorough evaluation of different design combinations.

Finally, the ability of an organism to be genetically modified and the availability of enzymes to perform desired biochemical transformations are necessary for successful implementation of proposed designs. Different organisms (and strains of an organism) have different inherent competencies for genetic transformation. This can range from some strains that are naturally transformable (they will readily take up free extracellular DNA) to strains that have no known method for genetic modification. In addition to having characterized enzymes that can perform the desired biochemical reaction, these two aspects are critical to tangibly implementing a desired biological design.

Interestingly, many of these same considerations have been discussed in the synthetic biology arena in considering the ideal attributes of a metabolic engineering host chassis (Jarboe et al. 2010). Some of the idea attributes for a metabolic engineering chassis were listed as:

1. Growth in mineral salts medium with inexpensive carbon sources
2. Utilization of hexose and pentose sugars
3. High metabolic rate
4. Simple fermentation process
5. Robust organism
6. Ease of genetic manipulation and genetic stability
7. Resistance to inhibitors
8. Tolerance to high substrate and product concentrations

6.2 Potential Design Starting Points

There are many different types of bacteria and each has evolved to thrive in a specific environment. This vast array of options is beneficial in choosing a host organism for producing biofuel because native properties can be seen as a head-start or work that does not have to be done. There are two main advantages that a host organism can provide. Either an organism can be chosen that already uses the feedstock efficiently or already produced the desired product. When testing the recombinant pathway to biofuel production it is beneficial to start with an organism such as *E. coli* that has a high growth rate, is genetically tractable, has a relatively well-defined system, and there are accurate computational models to predict production of metabolites. Producing biofuels from bacteria that have a sugar feedstock does not end up being very efficient in the overall life cycle analysis so the paths to produce the biofuel should be implemented into a bacteria with a lignocellulosic feedstock or a photosynthetic organism since it is more likely to be cost-efficient and sustainable. Terpenes can be produced using a secondary metabolic pathway that leads to isoprenoid production and can be modified to further produce long chain terpenes that are used for jet fuel or isopropanol which can be used similar to ethanol. When choosing which genes to integrate into the host organism it is beneficial to choose enzymes native to similar growing environments as the host cell.

Optimization techniques include plasmid copy number, codon optimization, promoter variation and overexpression, and reduction of competing pathways.

6.2.1 Organisms with Native Product Formation

One common and very effective strategy for microbial fuel production has been the employment of organisms which naturally produce the chemical of interest. As one prominent example, *Clostridium acetobutylicum* and *Clostridium beijerinckii* were both used significantly in the earlier part of the twentieth century in the acetone–butanol–ethanol (ABE) process.

Because these organisms already have the capability to produce and secrete ethanol and butanol, research efforts could focus on optimizing these species for biofuel production by removing alternative carbon by-products such as butyrate or acetone. Since the introduction of this process, engineering efforts have focused on improving product yield and specificity, broadening substrate range, and improving product tolerances. Very recently, significant advances in these areas have been achieved through the application of systems biology techniques. Both of these organisms exhibit a concerted metabolic shift from acidogenesis to solventogenesis during mid-late fermentation. Transcriptomic analysis of *C. beijerinckii* during that shift recently revealed the transcriptional regulatory changes that underpin this shift, as well as the physiological responses of sporulation and chemotaxis (Shi and Blaschek 2008). Another study used bioinformatic analysis of several clostridial genomes to predict potential small RNAs (sRNA) in these organisms (Chen et al. 2011). Future research could use these discoveries to control the regulatory shift to solventogenesis or to direct all solventogenesis to a desirable product, such as ethanol or butanol, thereby improving overall productivity and yield and reducing downstream separation costs. As another example, the ethanologen *Zymomonas mobilis* is naturally highly optimized for ethanol production, reaching near theoretical yields. Recent genomic, transcriptomic, and metabolomic experiments by the BioEnergy Science Center (BESC) at Oak Ridge National Labs have resulted in a flood of information about this organism and mutant strains which have already been used as a discovery tool to uncover the genetic mechanism for acetate tolerance (Yang et al. 2009a, b, 2010). This discovery will allow better productivity by *Z. mobilis* on hydrolysate from dilute acid pretreatment methods that use acetic acid, and, more generally, this demonstrates the utility of systems biology techniques for elucidating complex or emergent behaviors.

6.2.2 Organisms with Native Substrate Utilization

An alternative strategy for selecting a suitable platform organism for biofuel production is to choose species that can naturally utilize a broad range of abundant substrates such as lignocellulose or syngas. These organisms can then be manipulated to manufacture a desired product. Many of these organisms are opportunistic or highly specialized to specific environments, so the challenge for metabolic engineers is to improve process tolerance and product yields. This is best accomplished by studying molecular and cellular functions as a basis for subsequent design/engineering efforts.

An example of building knowledge and extrapolating it to design is the anaerobic, cellulolytic microorganism, *Clostridium thermocellum*. Microarray experiments on *C. thermocellum* have revealed substrate- and product-dependent transcriptional responses that will be valuable for improving ethanol yield. Quantitative proteomic (Raman et al. 2009) and lipidomic (Herrero et al. 1982) studies have examined its physiological responses to pretreatment inhibitors and high ethanol concentrations. Building upon these data, a constraint-based metabolic model of *C. thermocellum* has also been constructed and used as a platform for the integration and interpretation of global gene expression data (Roberts et al. 2010; Gowen and Fong 2010) This data-integrated computational model can then be used to predict genetic modifications that would increase the product yield and productivity of *C. thermocellum* for the production of either ethanol or hydrogen. These efforts are likely to rapidly expand to other cellulolytic and hemicellulolytic organisms, as well as to organisms that grow on syngas, as more genomic information becomes available.

The strategy of starting with a cellulolytic organism has the benefit of utilizing the diverse and effective cellulose systems native to these organisms, because these complex and coordinated enzyme systems would be difficult and costly to reproduce in laboratory model strains such as *Escherichia coli* and *Saccharomyces cerevisiae*. The challenge then is to overcome poor productivity and yield and to optimize process tolerance and, as described above, systems biology techniques will continue to direct and enable these efforts.

Every additional function a cell has consumes more energy and therefore leaves less energy for the process that is making the product. With this is mind it would be logical to look for a minimalistic cell that only carries the functions necessary to survive and to produce a desired product. Some research groups are already on the quest for a "minimal cell" or "generic host". This bare-bones cell only contains the essential genetic information required to maintain viability under certain conditions. A minimalist cell can be generic and open for inclusion of specific functions or it can be designed from scratch to perform a specific function. With the current knowledge of biological systems it is very possible that this theory of a minimalist cell would result in limited efficiency because the complexity of the cell is not completely understood.

6.3 The Systems and Synthetic Biology Complement

In other engineering disciplines (mechanical, electrical, civil, chemical, etc.) de novo design based upon theory/knowledge and desired function is common. The ability to successfully implement *de novo* design generally relies on two critical components, good characterization of available materials/building blocks/components and the ability to accurately physically implement the design from a blueprint or design specifications. In biological systems, both of these aspects have historically been hurdles that are now being overcome. The ability to effectively implement *de novo* biological designs enables a more straightforward approach to biological engineering.

The knowledge base and comprehensive characterization of fundamental biological components is born from the centuries of careful biological and biochemical experiments that have been conducted. Recent systems biology research has expedited the characterization of biological components and added additional information on coordinated activity and interactions.

Synthetic biology research has developed a variety of parallel approaches to the synthesis and assembly of DNA constructs of various lengths (up to a small genome). With the ability to directly synthesize or assemble genetic constructs to exactly correspond to a design, biological engineering may no longer have a significant hurdle at the construction step. Thus, if the knowledge base is comprehensive and well-curated and methods of implementation exist, biological engineering can focus on design.

As described, the recent contributions of systems biology and synthetic biology to biological engineering are highly complementary. The systems-wide analyses and modeling of systems biology help to provide the basis for cellular-level design. Details of this design can be translated to base-by-base specification for synthesis and implementation. Synthetic and molecular biology tools can be used to fine tune cellular-level designs and implement them into engineered strains. Engineered strains can then be evaluated and characterized for function by experimental systems biology techniques. If necessary, this process can be iterated for continuous improvement of function (Fig. 6.2).

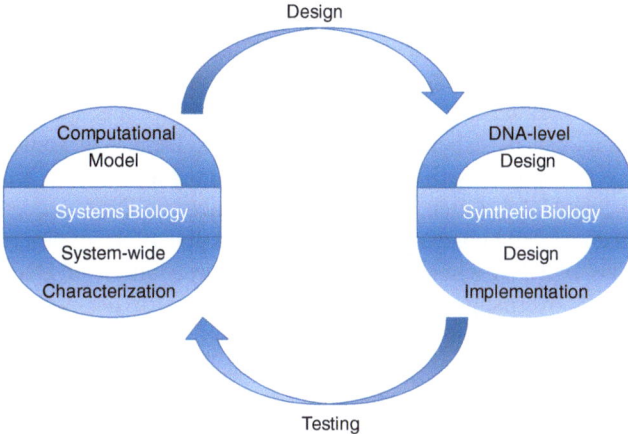

Fig. 6.2 Graphical representation of complementary aspects of systems biology and synthetic biology as applied to biological design and engineering

The collective biological knowledge for many organisms is now sufficient to attempt whole-cell manipulations/design. Currently, one of the most successful tools for conducting whole-cell designs is to use computational models, such as a constraint-based model (Chap. 4.2.1). After developing a constraint-based model for an organism of interest, there exist a number of different design algorithms that can be used to help to guide a whole-cell design. The most common of these algorithms are variants of optimization-based algorithms such as OptKnock (Burgard et al. 2003) that uses a bi-level optimization where the two objectives that are considered are typically your chemical of interest and cellular growth. More recent algorithms include OptForce (Ranganathan et al. 2010) and EMILiO (Yang et al. 2011) and evolution-based algorithms such as simulated annealing (Gonçalves et al. 2012). Algorithmic design using a computational model can help to generate whole-cell designs that suggest genetic modifications (gene additions, gene deletions, or changes in gene expression) and/or environmental conditions that would improve the desired cellular function.

The cellular-level design accomplished by constraint-based models and associated algorithms are broad and comprehensive at a cellular level, but do not necessarily contain all details necessary for direct experimental implementation. Especially, for cases where new genes are to be added to a system for heterologous expression, additional design specifications need to be made at the molecular/genetic level. In this case where heterologous expression of a gene is desired, there are numerous considerations to account for. Is the GC content and codon usage between the organisms (host chassis organism and the organism natively possessing the desired gene) sufficiently different that the gene should be codon optimized for the new chassis? What level of expression is desired for the target gene? Will the expression be controlled by designing promoters, ribosome binding sites, RNA secondary structure, or by plasmid copy number? Finally, if the desired gene and ancillary DNA sequences are to be synthesized, how are the DNA sequences going to be assembled and amplified? All of these questions can directly affect the function of the gene and can change the base-by-base detail of what will be experimentally implemented.

If all of the cellular-level and genetic-level designs are specified, then the designed strains can be experimentally constructed. Details of some of the synthetic biology methods used to assemble and integrated DNA sequences are described in Sect. 5.1.1. Often, the method by which the DNA sequences will be assembled is considered concurrent with the initial genetic-level design as the different methods of DNA assembly often require the addition of DNA sequences that are used as overhangs or recognition sequences for the assembly method. While DNA synthesis costs continue to decrease, there is often also decision about which assembly method to use based upon the length of the desired DNA sequence to by synthesized and the cost of synthesis (i.e., a single continuous DNA sequence of 2,000 bases can be directly synthesized reducing the need for DNA assembly, but it may be much more cost-effective to synthesize 20, 100-base long DNA fragments and assemble them).

Once the strains have been designed and experimentally constructed, all that remains is to evaluate the degree of functional success achieved. For chemical production applications, this typically involves two primary areas of analysis. The first is the use of some analytical chemistry analysis to quantitatively assess the amount of production of the desired chemical. A wide range of analytical chemistry techniques can be employed including HPLC (high performance liquid chromatography), mass spectrometry, NMR, or even assays. The second area of analysis that is typically conducted is some type of characterization of cellular function. In an ideal case, this would include some of the systems biology experimental techniques such as transcriptomics, proteomics, fluxomics, or metabolomics. System-wide evaluation of cellular function can provide insight into how effectively a proposed cellular design is being implemented in the in vivo setting. Furthermore, system-level measurements can be used as information and input to improve the function of an organism if it is not behaving as was predicted or desired (Gowen and Fong 2010).

6.4 Expanding the Options

With the above outlined biological engineering process, it is possible to specify a target chemical and organism and engineer the organism to produce the desired chemical. In this process, there are decisions that need to be made regarding what is the target chemical and most appropriate organism to use as the host organism. These decisions are most intelligently decided by understanding the chemical in question, the biological characteristics of different organisms, and the availability of different enzymes to implement the desired function. By necessity, these decisions are made based upon what information is available in each of these areas. Ongoing discoveries and research are continually adding to our chemical and biological knowledge to expand the possibilities of what can be attempted.

6.4.1 Bioprospecting

One of the biggest advantages of using biological processes is the immense diversity of biological function available. Bioprospecting is expanding the collective library of biological information in the scientific community by searching for and identifying novel organisms and functionalities. Systems biology provides tools for high throughput genetic and phenotypic characterization, thereby reducing the time and effort spent in the wet lab. As an example, bioprospecting for improved utilization of lignocellulose has turned up interesting and novel cellulolytic species from a variety of environments including terrestrial (Kim et al. 2009; Semêdo et al. 2004), aquatic (Distel et al. 2002; Podosokorskaya et al. 2011; Miroshnichenko et al. 2008) and ruminal settings (Chassard et al. 2012; Chang et al. 2011). In addition to testing for cellulolytic capabilities, some of the novel species possess additional interesting metabolic capabilities such as the ability to use carbon monoxide as a substrate (Bruant et al. 2010) or the ability to produce and accumulate high levels of oil (Araujo et al. 2011). With the development of lower cost, higher throughput DNA sequencing technologies, the field of metagenomics has made it possible to explore biological diversity at the genetic level, without the need to isolate or identify the originating host organism. For example, metagenomics can be used to search for novel cellulases (Li et al. 2009; Sommer et al. 2010) or as a means of studying chemical production capabilities, such as alkane production in cyanobacteria (Schirmer et al. 2010).

6.4.2 Metagenomics

While bioprospecting typically results in the isolation of intact novel organisms, it is possible to search and identify new biological information based upon genetic content alone. In this case, high throughput DNA sequencing is used to generate DNA sequence data for any genetic material (e.g., soil sample, air sample, etc.). Metagenomics is a type of high-throughput characterization that is rising in popularity because it tests a small ecosystem of organisms that grow and thrive together. Most current characterization techniques involve isolating a specific organism and allowing it to propagate for testing, but many species go unidentified in this manner. Metagenomics catches all of the organisms in the sample being tested. In this manner, it is possible to generate genetic information without the intact context of the host organism. Thus, it is possible to expand the range of possible enzyme functions without the requirement of knowing all of the details of the original host organism. Previously, this type of biological information may have been of limited use for applications, but growth of synthetic biology and DNA synthesis provides a natural avenue for implementing DNA information.

6.4.3 Bioinformatics

Research in systems biology has been underway for many years and until recently it has been a slow accumulation of knowledge obtained through numerous time-consuming laboratory experiments. While science will always need the creativity and innovation of its researchers, computers have proven to be very useful in a wide range of biological applications. The growth and compilation of biological knowledge has led to an analysis problem where often there is more information that can be intelligently deciphered. Bioinformatics has developed as a field to help address the analysis challenges that face biology.

One of the central components of bioinformatics is the use of computers in biology. A core component of bioinformatics is data mining where computational algorithms are used to study and analyze biological information. Most often these algorithms are reflections of hypotheses or current beliefs in biology and help to search through large data sets to not only test hypotheses, but also expand our knowledge by identifying novel attributes. In this manner, bioinformatics can help add to the biological knowledge further expanding possible design options.

Bioinformatic analyses cover a broad spectrum of biological research. This ranges from DNA sequence analyses to ecosystems or organism systems. For the purposes of biofuel applications, many of the most useful applications are function based such as analyses that link biochemistry and enzyme activity.

Chemical Enzyme databases have been developed to suggest enzyme pathways for chemical production and predict chemical compounds once exposed to certain enzymes. One database (BNICE) uses the thermodynamic data available for existing reactions to predict the most favorable pathway. By using the energies known for chemical bond formation, and the extent to which enzymes can reduce the energy needed, a change in Gibb's free energy (ΔG) is determined for each reaction and it is known that having a negative ΔG means that the reaction will happen spontaneously. The model will provide different pathway options using different enzymes and provide the ΔG for each proposed reaction pathway.

References

Araujo GS, Matos LJ, Gonçalves LR, Fernandes FA, Farias WR (2011) Bioprospecting for oil producing microalgal strains: evaluation of oil and biomass production for ten microalgal strains. Bioresour Technol 102(8):5248–5250. doi:10.1016/j.biortech

Bruant G, Lévesque MJ, Peter C, Guiot SR, Masson L (2010) Genomic analysis of carbon monoxide utilization and butanol production by Clostridium carboxidivorans strain P7. PLoS ONE 5(9):e13033. doi:10.1371/journal.pone.0013033

Burgard AP, Pharkya P, Maranas CD (2003) Optknock: a bilevel programming framework for identifying gene knockout strategies for microbial strain optimization. Biotechnol Bioeng 84(6):647–657. doi:10.1002/bit.10803

Chang L, Ding M, Bao L, Chen Y, Zhou J, Lu H (2011) Characterization of a bifunctional xylanase/endoglucanase from yak rumen microorganisms. Appl Microbiol Biotechnol 90(6): 1933–1942. doi:10.1007/s00253-011-3182-x

Chassard C, Delmas E, Robert C, Lawson PA, Bernalier-Donadille A (2012) Ruminococcus champanellensis sp. nov., a cellulose-degrading bacterium from human gut microbiota. Int J Syst Evol Microbiol 62(Pt 1):138–143. doi:10.1099/ijs.0.027375-0

Chen Y, Li K, Pu H, Wu T (2011) Corticosteroids for pneumonia. Cochrane database of systematic reviews issue 3. no: CD007720. doi:10.1002/14651858.CD007720.pub2

Distel DL, Morrill W, MacLaren-Toussaint N, Franks D, Waterbury J (2002) Teredinibacter turnerae gen. nov., sp. nov., a dinitrogen-fixing, cellulolytic, endosymbiotic gamma-proteobacterium isolated from the gills of wood-boring molluscs (Bivalvia: Teredinidae). Int J Syst Evol Microbiol 52(Pt 6):2261–2269. doi:10.1099/ijs.0.02184-0

Gonçalves E, Pereira R, Rocha I, Rocha M (2012) Optimization approaches for the in silico discovery of optimal targets for gene over/underexpression. J Comput Biol 19(2):102–114. doi:10.1089/cmb.2011.0265

Gowen CM, Fong SS (2010) Genome-scale metabolic model integrated with RNAseq data to identify metabolic states of Clostridium thermocellum. Biotechnol J 5(7):759–767. doi:10.1002/biot.201000084

Herrero AA, Gomez RF, Roberts MF (1982) Ethanol-induced changes in the membrane lipid composition of Clostridium thermocellum. Biochim Biophys Acta 693(1):195–204. doi:10.1016/0005-2736(82)90487-4

Jarboe LR, Zhang X, Wang X, Moore JC, Shanmugam KT, Ingram LO (2010) Metabolic engineering for production of biorenewable fuels and chemicals: contributions of synthetic biology. J Biomed Biotechnol 2010:761042. doi:10.1155/2010/761042

Kim BC, Lee KH, Kim MN, Kim EM, Min SR, Kim HS, Shin KS (2009) Paenibacillus pini sp. nov., a cellulolytic bacterium isolated from the rhizosphere of pine tree. J Microbiol 47(6):699–704. doi:10.1007/s12275-009-0343-z

Li LL, McCorkle SR, Monchy S, Taghavi S, van der Lelie D (2009) Bioprospecting metagenomes: glycosyl hydrolases for converting biomass. Biotechnol Biofuels 2:10. doi:10.1186/1754-6834-2-10

Miroshnichenko ML, Kublanov IV, Kostrikina NA, Tourova TP, Kolganova TV, Birkeland NK, Bonch-Osmolovskaya EA (2008) Caldicellulosiruptor kronotskyensis sp. nov. and Caldicellulosiruptor hydrothermalis sp. nov., two extremely thermophilic, cellulolytic, anaerobic bacteria from Kamchatka thermal springs. Int J Syst Evol Microbiol 58(Pt 6):1492–1496. doi:10.1099/ijs.0.65236-0

Podosokorskaya OA, Kublanov IV, Reysenbach AL, Kolganova TV, Bonch-Osmolovskaya EA (2011) Thermosipho affectus sp. nov., a thermophilic, anaerobic, cellulolytic bacterium isolated from a Mid-Atlantic Ridge hydrothermal vent. Int J Syst Evol Microbiol 61(Pt 5):1160–1164. doi:10.1099/ijs.0.025197-0

Raman B, Pan C, Hurst GB, Rodriguez M Jr, McKeown CK, Lankford PK, Samatova NF, Mielenz JR (2009) Impact of pretreated Switchgrass and biomass carbohydrates on clostridium thermocellum ATCC 27405 cellulosome composition: a quantitative proteomic analysis. PLoS ONE 4(4):e5271. doi:10.1371/journal.pone.0005271

Ranganathan S, Suthers PF, Maranas CD (2010) OptForce: an optimization procedure for identifying all genetic manipulations leading to targeted overproductions. PLoS Comput Biol 6(4):e1000744. doi:10.1371/journal.pcbi.1000744

Roberts SB, Gowen CM, Brooks JP, Fong SS (2010) Genome-scale metabolic analysis of Clostridium thermocellum for bioethanol production. BMC Syst Biol 4:31. doi:10.1186/1752-0509-4-31

Semêdo LT, Gomes RC, Linhares AA, Duarte GF, Nascimento RP, Rosado AS, Margis-Pinheiro M, Margis R, Silva KR, Alviano CS, Manfio GP, Soares RM, Linhares LF, Coelho RR (2004) Streptomyces drozdowiczii sp. nov., a novel cellulolytic streptomycete from soil in Brazil. Int J Syst Evol Microbiol 54(Pt 4):1323–1328. doi:10.1099/ijs.0.02844-0

Schirmer A, Rude MA, Li X, Popova E, del Cardayre SB (2010) Microbial biosynthesis of alkanes. Science 329(5991):559–562. doi:10.1126/science.1187936

Shi Z, Blaschek HP (2008) Transcriptional analysis of clostridium beijerinckii NCIMB 8052 and the hyper-butanol-producing mutant BA101 during the shift from acidogenesis to solventogenesis. Appl Environmental Microbiol 74(24):7709–7714. doi:10.1128/AEM.01948-08

Sommer MO, Church GM, Dantas G (2010) A functional metagenomic approach for expand-
    ing the synthetic biology toolbox for biomass conversion. Mol Syst Biol 6:360. doi:10.1038/
    msb.2010.16
Yang L, Cluett WR, Mahadevan R (2011) EMILiO: a fast algorithm for genome-scale strain
    design. Metab Eng 13(3):272–281. doi:10.1016/j.ymben
Yang S, Pappas KM, Hauser LJ, Land ML, Chen GL, Hurst GB, Pan C, Kouvelis VN, Typas
    MA, Pelletier DA, Klingeman DM, Chang YJ, Samatova NF, Brown SD (2009a) Improved
    genome annotation for Zymomonas mobilis. Nat Biotechnol 27(10):893–894. doi:10.1038/
    nbt1009-893
Yang S, Tschaplinski TJ, Engle NL, Carroll SL, Martin SL, Davison BH, Palumbo AV, Rodri-
    guez M Jr, Brown SD (2009b) Transcriptomic and metabolomic profiling of Zymo-
    monas mobilis during aerobic and anaerobic fermentations. BMC Genomics 10:34.
    doi:10.1186/1471-2164-10-34
Yang S, Land ML, Klingeman DM, Pelletier DA, Lu TY, Martin SL, Guo HB, Smith JC, Brown
    SD (2010) Paradigm for industrial strain improvement identifies sodium acetate toler-
    ance loci in Zymomonas mobilis and Saccharomyces cerevisiae. Proc Natl Acad Sci USA
    107(23):103

Chapter 7
Building Engineered Strains

Abstract A critical step in engineering design is the ability to accurately construct something to directly translate a conceptual design to practice. Due to the relatively short history of molecular biology and the limitations in knowledge and technology, biological engineering has been hindered by a lack of ability to easily implement conceptual designs. Recent developments have attempted to standardize different aspects of biology to facilitate and expedite biological engineering. Standardization of genetic parts will not only ease the methodological hurdles for biological engineering, but it will also enable biological engineering to focus more exclusively on conceptual design of function rather than being constrained by practical limitations.

A classic hallmark of humans is the ability to conceive, build, and use tools to achieve desired goals. Advances in construction methods relevant for most disciplines allow building or construction to occur quickly and with minimal thought. For example, humans have been working with iron for more than 3000 years and mass-produced steel for more than 160 years. This established history of metallurgy allows for devices to be constructed out of metal quickly and efficiently. In terms of the overall process of biological process development, this chapter will focus on the specific aspect of building/construction as it is one of the major hurdles for biological engineering.

The ability to quickly and efficiently construct desired devices has not been possible for biological systems, especially if considering a functional cell to be the device to be constructed. Contrary to the long history and established methodologies in other disciplines, scientists have only been working in detail with the fundamental material of biological systems (DNA) for a little over 50 years. While great advances have been made in a short period of time, there are still fundamental pieces of information that are still being discovered, such as the debated possibility of growing and incorporating arsenic instead of phosphorus into biological building blocks [including DNA (Wolfe-Simon et al. 2011)].

The shorter history of working with DNA in biological systems leads to very practical limitations in terms of what can be achieved in building biological systems. In most engineering disciplines other than biological engineering

S. M. Clay and S. S. Fong, *Developing Biofuel Bioprocesses Using Systems and Synthetic Biology*, SpringerBriefs in Systems Biology, DOI: 10.1007/978-1-4614-5580-6_7, © The Author(s) 2013

(mechanical, electrical, chemical, computer, civil), the historical foundations of those fields have provided sufficient depth of knowledge and tools to prospectively design and physically implement a product (Table 7.1). In the more established engineering disciplines, the different aspects of knowledge, design, and construction are typically segregated and discrete steps. The implication of this is that each of the steps can be addressed independently without major concerns for the other downstream steps. For example, an internal combustion engine can be designed based upon principles of combustion and mechanics. Design efforts produce a blueprint/schematic of the proposed engine based upon theory and desired functionality. The design step can be considered almost exclusively based upon the fundamental design principles and desired function without much regard to the actual construction of the engine. The techniques and capability to construct the engine are assumed as methodologies for metal working and machining are available.

The ability to segregate knowledge, design, and construction into discrete, independent steps has not always been possible for biological systems and biological engineering. Often all three of these components are interwoven due to different limitations. The design process often is limited by the amount of information/ knowledge available and sometimes, biological design cannot occur due to a lack of information. For example, an organism can be engineered to produce specific polyketides through genetic engineering, but this ability is contingent upon having an understanding of polyketide synthases. In other areas such as terpenoid production, the design process is not as straightforward as the level of knowledge for terpene synthases is not as well established as for polyketide synthases. One current example of this is the desire to heterologously express and produce the terpenoid paclitaxel (Taxol) for cancer treatment, but the complete biosynthetic pathway has not yet been elucidated. Thus, an engineered strain for bioproduction of paclitaxel is not yet possible.

Of specific concern for this section, there have also been limitations in construction that influence the design step. Currently, the single most commonly used approach for implementing biological designs is the use of genetic engineering methodologies. Thus, there is really no consideration on what the materials of construction will be (DNA), but the ability to exactly build something to match a *de novo* design specification and to have it compatible with other components by a set of standards has been problematic. Practically speaking what this has led to in biological engineering is a design process where designs are constrained by what information is known and also how possible it is implement a proposed design. Until recently, this has caused the engineering process for biological systems to be a process of continual feedback and iteration rather than a linear building/construction process.

In terms of building engineered strains, recent developments (mainly associated with technology and methodology improvements) have sought to address the constraints and limitations associated with the implementation or construction of genetic constructs. The developments have taken different approaches to address the two main problems: (1) the ability to construct something that exactly matches the blueprint/design and (2) standardization to facilitate compatibility.

Table 7.1 Examples of aspects of design in various engineering disciplines

Engineering Discipline	Goal	Theory	Parts	Product
Electrical	Electronic device	Kirchoff's law Ohm's law	Transistors, resistors, capacitors, inductors, diodes	Circuit board
Chemical	Chemical process	Conservation of mass and energy, reaction kinetics	Catalysts, chemicals	Chemical reactor
Biological	Biological process	Central dogma of molecular biology	DNA	Cell

7.1 Standardization of DNA

Standardization has many different facets but the two that we will focus on here are the standardization of requirements and the standardization of part interoperability. When mentioning the standardization of requirements, what is meant is a set of guidelines or rules that are accepted in the field and must be abided by. For standardization of part interoperability, the intention is to facilitate the actual construction and implementation during the actual building process.

Standardization of requirements occurs in all engineering disciplines and impacts various facets of that field. Every field has some form of standards that related to safety and ethics. In addition, there are also technical specifications/standards that are established. For biological systems, these standards in the areas of ethics and technical specifications are often unspoken and can vary from lab to lab.

In the area of safety and ethics, some of the guidelines are well established whereas others are almost left to the individual to decide. Safety standards are relatively uniform, especially for work that is conducted in an academic setting or sponsored by federal funding. For example, the National Institutes of Health have established, published guidelines regarding research using recombinant DNA (http://oba.od.nih.gov/rdna/nih_guidelines_oba.html). One of the challenges is that methodologies and capabilities are continually changing and thus, policy guidelines must also evolve. This is demonstrated by the discussion on amending the recombinant DNA guidelines to account for synthetic nucleic acids (http://oba.od.nih.gov/rdna_rac/rac_pub_con.html).

Ethical considerations are less well-defined than safety in terms of policies and guidelines. The ethical implications of genetic engineering and designed organisms are more grounded in personal world views rather than technical detail. Some of the more widely debated related subject areas are genetically modified foods and whole organism cloning. Specific to the topic of biofuel production and synthetic or recombinant DNA are questions regarding the scope of genetic interventions. Are we ethically comfortable with modifying a single gene in an organism? An entire pathway? The entire organism? As mentioned previously, the technical capability exists to synthetically create the DNA for an organism's entire genome. If an entire genome is synthetically created and no known organism's genome is used as template, would the new genome and related organism be a synthetically created new life form? How much modification to an existing organism would need to be made to designate a new species? The answers to these questions and how the scientific community should proceed are pressing issues that need to be addressed.

The other major area of standardization that is being addressed is the standardization of technical specifications. This is largely a practical consideration and follows from the concept of interchangeable parts. A nut and bolt combination works well when there are standardized threads (depth and pitch) that match between the nut and bolt. Furthermore, if depth and pitch of the thread is standardized, replacement nuts can be easily found for a given bolt if a nut should be lost. This type of technical standardization is common in most fields where something is built or constructed.

Biological research has long been one where standardization is not common. Polymerase chain reaction (PCR) is one of the most useful and prevalent methods used to amplify DNA, but every individual DNA sequence to be amplified requires the design and construction of unique DNA primer sequences to be used in the PCR reaction. Furthermore, depending upon the characteristics of the DNA sequence to be amplified, changes may need to be made to the actual experimental protocol of the method. The challenges associated with the uniqueness of biology are pervasive and can be seen from designing individual probes for gene expression microarrays to having mass and fragmentation patterns for mass spectrometry applications.

In terms of genetic engineering, one of the most comprehensive attempts to establish standardized technical specifications for DNA is the BioBrick formalism (Shetty et al. 2008). The BioBrick concept establishes a standard format for all DNA sequences where a sequence of interest is flanked upstream and downstream by standard DNA sequences. The added DNA sequences are cut sites that are recognized by specific restriction enzymes. In this format, the desired DNA sequence is flanked upstream by the restriction enzyme recognition cut sites for EcoRI and XbaI and downstream by the restriction enzyme recognition cut sites for SpeI and PstI (Fig. 7.1). Using this formalism, the desired DNA sequence (DNA part) can be of any length with any sequence (as long as the sequence does not contain cut sites for EcoRI, XbaI, SpeI, or PstI). Different DNA sequences that have this format can be manipulated by the same protocol by using the four restriction enzymes EcoRI, XbaI, SpeI, and PstI. Currently, thousands of DNA sequences of varying length and function exist in a centralized DNA repository (www.partsregistry.org) and all of these parts can be worked with using standardized protocols (contrast this with having to design and synthesize primers for each sequence individually).

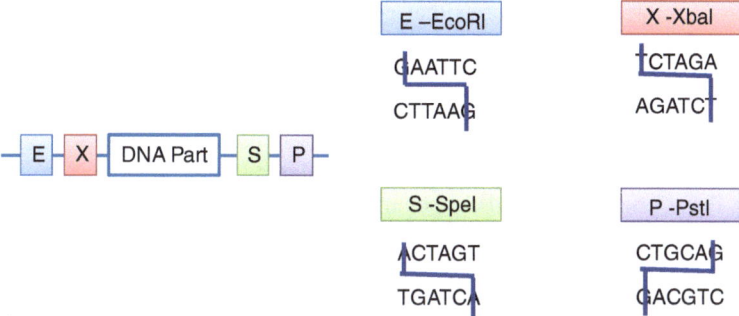

Fig. 7.1 Graphical depiction of the BioBrick format for standardizing DNA parts. A target sequence (DNA part) is flanked upstream by restriction enzyme cut sites EcoRI (E) and XbaI (X) and downstream by SpeI (S) and PstI (P)

When using the BioBrick format, DNA sequences can be isolated and assembled with other DNA sequences quickly by cutting the sequence using restriction enzymes and ligating the desired sequences together using DNA ligase. This can be done for as many DNA sequences as desired and in any order that is desired. Furthermore, there exist a number of DNA plasmids that have been constructed with replication origins and different antibiotic resistance genes as markers. Thus, the same methodologies used to assemble DNA sequences together can be used to generate a self-replicating plasmid housing DNA sequences of interest.

7.2 Interoperability of DNA Constructs

The challenges associated with building or constructing biological systems continues beyond developing methods to standardize genetic parts. One of the characteristics of biological systems is that the components in an organism are all highly connected (metabolic network, regulatory network, protein interaction network). Thus, even after being able to construct a desired DNA sequence properly, there is a challenge in having the DNA sequence expressed and functional within the network context of existing components. This can be viewed as a challenge of interoperability or compatibility.

In terms of controlling expression of the designed DNA sequence, various tools are available to help to control or to dictate the expression level of an introduced or modified DNA sequence. Fundamentally, expression is controlled at the transcription and translation steps by the pairings of promoter and DNA polymerase for transcription and ribosome binding site (RBS) and ribosome for translation. The level of transcription can be altered by modifying the binding strength between a promoter and DNA polymerase. To date, the most effective means of achieving this has been accomplished empirically by developing and characterizing promoter libraries that contain promoters with small variations in sequence. The level of translation can be similarly altered by modifying the interaction between the RBS and the ribosome. In this case, the interaction between the RBS and ribosome follows nucleic acid pairing, so predictions on the strength of this interaction can be made to help to guide design of this interaction (an example of this is the ribosome binding site calculator (Salis et al. 2009)). In addition to these primary methods, other methodologies such as codon optimization and designing RNA hairpins can influence the expression levels of a target DNA sequence.

Even when a DNA sequence is designed, constructed, and care is taken to try to control its expression, the DNA sequence may not function as desired within the context of a living organism. In these instances, the ability to efficiently build and modify a biological system is severely limited by the state of our knowledge. It is often difficult to predict all of the downstream consequences of a genetic modification and in some instances there may be no means of predicting how an introduced gene/protein (for example) will interact with other existing genes/proteins. In these instances, the most common approach to address interoperability issues is an empirical approach where molecular evolution is used. Variants of the desired gene/protein are generated and screened in context for the desired function. Currently, the use of molecular evolution is often necessary for implementation of even small genetic constructs that contain only a couple genes.

References

Salis HM, Mirsky EA, Voigt CA (2009) Automated design of synthetic ribosome binding sites to control protein expression. Nat Biotechnol 27(10):946–950. doi:10.1038/nbt.1568

Shetty RP, Endy D, Knight TF Jr (2008) Engineering BioBrick vectors from BioBrick parts. J Biol Eng 2:5. doi:10.1186/1754-1611-2-5

Wolfe-Simon F, Switzer Blum J, Kulp TR, Gordon GW, Hoeft SE, Pett-Ridge J, Stolz JF, Webb SM, Weber PK, Davies PC, Anbar AD, Oremland RS (2011) A bacterium that can grow by using arsenic instead of phosphorus. Science 332(6034):1163–1166. doi:10.1126/science.1197258

Chapter 8
State of the Field and Future Prospects

Abstract Biofuel production has the potential to have a great and lasting impact on the society and is a focal point of biological and metabolic engineering research. There have been great advances in the ability to engineer biological systems, and it can be seen that there lies many possibilities to expand the use of biology in producing a variety of chemicals, not just biofuels.

The chemical field has started taking advantage of biological systems to make commodity chemicals as well as fine chemicals due to their much lower processing costs. So far the number of chemicals made by biological means is limited mostly to existing metabolites that are more easily purified and to a few fine chemicals that have been the subject of extensive research.

We are really just seeing the beginning of chemistry and synthetic biology coming together in a mutually beneficial way that will provide a sustainable future for many different industries. Chemists are realizing that using organisms as factories that require few inputs such as sugars or sunlight provides a much more cost-effective process that reduces the hazards involved with disposing of mass quantities of chemical waste and leaving a much larger carbon footprint.

Chemists are starting to see that the way in which organisms have evolved has great significance and generally exposes the most energy-efficient route to producing chemicals.

Elucidating the pathways to producing an unlimited variety of chemicals using organisms lies in the computational framework being established. Once there is a complete database of enzymes and how they affect any different chemical, it will be possible to use any combination of enzymes to produce an endless number of chemicals. While these enzymes could work outside of the organism, it may prove to be more beneficial to incorporate the process in the organism. To choose which organism would be most suited for the process, data as to which enzymes are in each organism, and at what efficiency recombinant enzymes work would need to be included in the database. The alternative to this method would be the success of a "minimal cell" that could accept all of the genetic information needed to carry out the production of a desired chemical.

S. M. Clay and S. S. Fong, *Developing Biofuel Bioprocesses Using Systems and Synthetic Biology*, SpringerBriefs in Systems Biology, DOI: 10.1007/978-1-4614-5580-6_8, © The Author(s) 2013

In order for biofuels to be integrated into our current infrastructure and life-style, there will need to be a transition period where different combinations of biofuels fulfill the supply. A biofuel that can be blended into gasoline or easily implemented at existing gas stations that can work in our current engines would be the easiest transition but blending anything other than ethanol at this point is not cost-effective.

Algal biodiesel seems to be a solution that would work well in terms of production, greenhouse gas emissions, and sustainability, but this would require people to have diesel engines and they are not currently popular in personal vehicles. Once the price of algal biodiesel is low enough to compete with gasoline, the benefits of it should outweigh gasoline enough to where people start transitioning to diesel engines to accommodate this and other biofuels which work better in diesel engines.

Ethanol as a main biofuel has already been implemented in Brazil where they use almost a quarter of the world's ethanol fuel supply. Ethanol is already being blended into gasoline here in the US by up to 10 % ethanol, but there is still some debate as to whether or not ethanol is harming our current gasoline engines. Modifications can be made to existing engines that would allow for pure ethanol usage in personal automobile engines or they are already manufacturing flex-fuel vehicles that are already designed to run with high levels of ethanol. As of now most ethanol in the US is being produced from corn, but it will only succeed as a major contributor to our future in biofuels if we can produce ethanol in cost-efficient ways that do not take up arable land and which reduce greenhouse gas emissions. Producing ethanol from cellulosic feedstock or cyanobacteria would be the best routes if ethanol is to become a mainstay in biofuels in the US, but producing propanol or butanol may prove to be a better option.

Propanol or butanol production from cellulosic feedstock or a photosynthetic organism would be a much better option in terms of fuel efficiency but there is much research to be done to develop a process that is efficient enough to become cost-effective when run on a large scale. Most likely these types of fuels will not be able to stand on their own and will be blended with ethanol, gasoline, or another type of fuel to provide energy efficiency and water resistance without raising the cost of the fuel too much.The future of sustainable biofuels will be a blend of many techniques initially and may never be reduced to one single biofuel that shines above the rest in terms of cost, efficiency, sustainability, and availability.

The application of new knowledge and techniques derived from systems biology and synthetic biology is leading to new approaches to biological engineering. The advances make it possible for engineering design approaches to be taken in a biological setting where the individual steps of the design-build-test paradigm can be treated somewhat independently. This allows for a new degree of intellectual freedom where design is constrained by creative limitations, not logistical construction limitations. Moving forward, as the possibilities for biological engineering broaden, more prospective thought must be given to macroscopic decision making to help to identify promising avenues of research priority.